ANALYTICAL CHEMISTRY

The Periodic Table of the Elements (simplified)

The Atomic Number is at top left; the Mean Atomic Mass is given below, to two decimal places. The trans-uranic elements are not included

The 'd-block' elements and Nos 57–71, lanthanum and the 'f-block' elements, sometimes called the 'rare earths'.

Nos 89–92 are sometimes grouped with actinium in a similar way to the 'f-block' elements

Group 1	Group 2											Group 3	Group 4	Group 5	Group 6	Group 7	Group 0
1 H 1.01																	2 He 4.00
3 Li 6.94	4 Be 9.01											5 B 10.81	6 C 12.01	7 N 14.01	8 O 16.00	9 F 19.00	10 Ne 20.18
11 Na 22.99	12 Mg 24.31											13 Al 26.98	14 Si 28.09	15 P 30.97	16 S 32.06	17 Cl 35.45	18 Ar 39.95
19 K 39.10	20 Ca 40.08	21 Sc 44.96	22 Ti 47.88	23 V 50.94	24 Cr 52.00	25 Mn 54.94	26 Fe 55.85	27 Co 58.93	28 Ni 58.69	29 Cu 63.55	30 Zn 65.38	31 Ga 69.72	32 Ge 72.59	33 As 74.92	34 Se 78.96	35 Br 79.90	36 Kr 83.80
37 Rb 85.47	38 Sr 87.62	39 Y 88.91	40 Zr 91.22	41 Nb 92.91	42 Mo 95.94	43 Tc 98	44 Ru 101.07	45 Rh 102.91	46 Pd 106.4	47 Ag 107.87	48 Cd 112.41	49 In 114.82	50 Sn 118.69	51 Sb 121.75	52 Te 127.60	53 I 126.90	54 Xe 131.30
55 Cs 132.91	56 Ba 137.33	57–71 La, etc. –	72 Hf 178.49	73 Ta 180.95	74 W 183.85	75 Re 186.21	76 Os 190.2	77 Ir 192.22	78 Pt 195.08	79 Au 196.97	80 Hg 200.59	81 Tl 204.38	82 Pb 207.20	83 Bi 208.98	84 Po 209	85 At 210	86 Rn 222
87 Fr 223	88 Ra 226.03	89 Ac 227.03	90 Th 232.04	91 Pa 231.04	92 U 238.03												

ANALYTICAL CHEMISTRY

An Introduction

Gerald F. Lewis, CChem, FRSC

SECOND EDITION

MACMILLAN

First edition published 1973 by
BDH Chemicals Ltd
Poole, BH12 4NN

Second edition published 1985 by
Higher and Further Education Division
MACMILLAN PUBLISHERS LTD
Houndmills, Basingstoke, Hampshire RG21 2XS
and London
Companies and representatives
throughout the world

Printed in Hong Kong

British Library Cataloguing in Publication Data
Lewis, Gerald F.
Analytical chemistry : an introduction.—
2nd ed.
1. Chemistry, Analytic
I. Title
543   QD75.2
ISBN 0-333-38567-5

CONTENTS

FOREWORD TO THE FIRST EDITION

A good basic knowledge of analytical chemistry procedures can be considered as being essential for most laboratory technicians irrespective of the type of laboratory work on which they are engaged. Many young men and women entering the chemical industry straight from school neither have the necessary practical experience nor sufficient theoretical background and most companies find it necessary to give these young entrants some form of intensive training course. In fact the Chemical and Allied Industries Training Board emphasises that such a course is essential and have produced outlines of suitable programmes. Courses developed in this way usually consist of practical laboratory work to familiarise the assistant with the processes and raw materials used and products manufactured by the particular company. They generally have a bias towards analytical chemistry since this discipline is most adaptable to meet the training needs. Unfortunately such practical based courses often provide only a very sketchy treatment of the theoretical aspects of the work.

This book is written by a chemist with a sound industrial background and based upon many years' experience of training young laboratory technicians. It combines a description of simple laboratory techniques and experiments with a sound theoretical backing at the right level for the young intake and as such fills a real gap in their training needs. I have no doubt that it will serve a most useful purpose in supplying the supplementary reading for technicians during their apprenticeship and will provide a useful book of reference to back up their subsequent general laboratory work.

C. WHALLEY
President – Analytical Division, Chemical Society
President – Society for Analytical Chemistry
[1973]

PREFACE TO THE FIRST EDITION

This book was written originally for the benefit of GCE 'O' and 'A' level new entrants to the Analytical Department of BDH Chemicals Ltd to be read during their initial training period. The purpose was to provide some basic theoretical background to the analytical methods which they were practising.

Having extended and refined the text, the author hopes that others will also find it useful, for example sixth formers preparing for GCE 'A' level Chemistry.

There has been no attempt to deal in depth with the various topics, the idea being to give the essential basic principles of a wide range of analytical techniques and to present, as it were, a summary of analytical chemistry as a whole. The individual subjects may be pursued in greater detail by reference to standard text-books and a small selection of examples for further reading is given in the bibliography.

The author hopes that the book will be easy to read and readily understood. In most industrial laboratories the day-to-day matters of practice and administration generally occupy the whole of the working day and sometimes more besides. On the occasion when one is able to refresh one's basic knowledge, it is useful to have at hand a set of condensed notes in which a large area can be scanned in a short time. In this sense it is hoped that the book will be of value to the more senior analyst from time to time. The author would like to thank Professor T. S. West of Imperial College of Science and Technology and Mr G. B. Thackray, B.Sc., M.Chem.A., F.R.I.C., Public Analyst for the City of

Portsmouth, who very kindly read the first draft and offered some valuable suggestions; also Mr C. Whalley, B.Sc., F.R.I.C., President of the Analytical Division, Chemical Society, for writing the Foreword.

G. F. Lewis
April 1973

PREFACE TO THE SECOND EDITION

This book was first published in April 1973 for the benefit of new entrants to the Analytical Department of BDH Chemicals Ltd. All of these young people were well qualified at the GCE 'O' level stage, but few had had much experience of practical analytical chemistry. A laboratory training scheme was in operation and the book was designed to supplement this by providing a basic theoretical background.

Since that time, the methodology of analytical chemistry has developed considerably. This new edition includes references to some techniques which were scarcely known 10 years ago. The opportunity has also been taken to improve the presentation of the original text.

In addition to its original intention, the first edition found a place in schools and colleges as a simple reference book. The author hopes that the new edition will continue to attract this interest.

There has been no attempt to deal in depth with the various topics. The basic principles are presented as simply as possible consistent with correct understanding, and given in a form which is both easy to read and to assimilate.

I wish to thank Dr E. J. Newman who kindly read the typescript and suggested a number of refinements. These have been incorporated in this new edition. Many thanks are also due to my colleagues Derek Moore and Christopher Thorpe who devised and produced the colour illustrations.

Poole, 1985   G. F. L.

1

INTRODUCTION

Chemistry is that area of science which is concerned with the study of the three states of matter — solid, liquid and gas. Analytical chemistry is the experimental study of the composition of solids, liquids and gases.

It may be that the identity of the material in question is unknown. In this case, the analyst must establish its identity before he goes any further. This is called *qualitative analysis*. Alternatively the material may be an impure example of a known chemical and the requirement is to identify and measure the impurities. Then again, the substance could be a complex mixture such as a crude oil or a metallic ore. Specialised methods of separation will then be needed to separate, identify and measure the components. These are different kinds of *quantitative analysis*.

Many different kinds of procedures and tests are available to the analyst to deal with these kinds of situation.

First we can gain a surprising amount of information by simply inspecting the material, and carrying out one or two simple tests. For example a liquid with an aromatic smell which evaporates leaving no residue is almost certainly *organic*. Alternatively, a white crystalline solid which melts at a high temperature and then solidifies on cooling, is very likely a metallic *inorganic* salt. This basic information will help the analyst to decide on his approach to the analysis proper.

Then we have the physical constants, for example, melting point, boiling point, refractive index and density. Every single chemical species gives specific values under defined experimental conditions. These measurements may then be used either to identify a material or one of its separated components, or to detect a deviation from one hundred per cent purity in a single chemical species.

Methods of analysis based upon the measurement of chemical reactions have their origins in antiquity. Classical methods, as they are called, still provide the foundation for the work of many industrial laboratories. These procedures often provide the methods of choice for determining the percentage or concentration of a major component with a high degree of precision. To these we may add a bewildering range of instrumental methods. These have greatly extended the capabilities of analytical chemistry in recent years. For example, it is a comparatively easy task to measure most elemental impurities at levels of a few parts per million with a high accuracy. A large group of modern methods depends on the absorption or emission of light by the molecules or atoms of the sample material under various conditions. These are *spectroscopic* methods. In another group, the sample, in solution or vapour condition, is passed through a specially packed column in which the components are separated prior to identification. These are the *chromatographic* procedures.

The results of a quantitative analysis may be expressed in a variety of ways. For example, the component may be quoted as a percentage, as a weight per unit volume or as parts per million.

It is important to appreciate that it is impossible to establish the absolute truth about the composition of a material. We may only estimate the composition as precisely and as accurately as possible. The results are subject to certain 'confidence limits' (see Chapter 11).

Furthermore, a single analysis on a single sample cannot provide any information regarding the average composition through the bulk of the material. To do so requires the analysis of a series of replicate samples taken from different parts of the bulk. It will then be possible to talk about the results with defined reservations.

2

SAMPLING

The first step towards achieving a meaningful analysis is to obtain an analytical sample which truly represents the bulk of the material. If this is not done, the results will be misleading or even useless, and careful laboratory work will have been carried out in vain.

There are some material types which do not present serious problems. These include liquids and solutions which are known to be single phase and of homogeneous composition. Even here, we have to guard against the presence of an immiscible impurity, or the possibility of the non-homogeneous distribution of the components of a solution.

In general, sampling difficulties increase with the size of the bulk of the material, its particle size and how far it differs from a single chemical species. It is often necessary to devise elaborate procedures, for example, in the case of several wagon loads of metallic ore.

Liquids are often sampled 'on line'. Where this is not feasible, a *sampling pipette* made of glass, polythene or stainless steel is inserted into the container, which might be a steel drum. A suitable volume of liquid is withdrawn and transferred to a labelled clean glass bottle for the attention of the laboratory.

Solids are often sampled with a *sampling spear*. This is a hollow tube shaped so that it may be inserted deep into the bulk of the material for the withdrawal of a sample. Frequently, several or many samples are taken, and either combined and mixed to give an average set of results or analysed separately, when the variation in composition within the batch may be judged.

Gases, which are almost always present at above atmospheric pressure, are often passed through a reducing valve to reduce the pressure to just above atmospheric. The gas may then be passed through a container until all the air has been displaced, when the container is sealed. A sample is then withdrawn for analysis by displacing the gas with a suitable inert liquid.

Impurities are not always distributed uniformly through a solid. They may be adsorbed on the surface and held in crystal imperfections. Solid laboratory samples should therefore always be ground to a fine powder unless their physical characteristics preclude this treatment. Failure to do this may again give rise to misleading results.

Special care needs to be taken to protect certain types of sample from changing in the laboratory pending the analysis. For example, hygroscopic materials must be adequately protected from a moist atmosphere.

The sampling step is as important as the analysis and should be undertaken by people who fully understand its implications.

3

THE BALANCE

The balance (see Plate 1) is the oldest and most basic instrument in the analytical laboratory. It also provides by far the most accurate and precise method of measurement. Its uses include:

Weighing samples.
Weighing gravimetric precipitates.
Weighing substances for preparing standard solutions.

The most commonly used balance may be used to weigh objects up to 200 g with an accuracy and precision of 0.0001 g. This means that three decimal places of a gram are accurate, but the fourth place is subject to an inherent error of ± 1.

When very small weights have to be measured, a 'semi-micro balance' or a 'micro balance' is required if the desired accuracy and precision is to be achieved. When in good working order, these balances perform respectively with accuracies and precisions of ± 0.00001 g (10 μg) and ± 0.000001 g (1 μg).

For less accurate work, for example, in the preparation of reagent solutions, 'top loading' balances are frequently used.

Earlier balances were 'free swinging' and had to be balanced about a central 'null point'. Today's balances give a direct reading, the beam coming to rest at a point between its two extremes. This is achieved by the use of damping cylinders acting on the ends of the beam. The weights, which are ring-form, are placed on the beam by means of a dial-operated system of levers. This avoids the need to handle them and thus altering their values.

5

Most balances are constant load instruments. One end of the beam carries a fixed load of, for example, 200 g. The other end carries the dial-operated weights. When the balance is empty, all the weights are in position. The object is placed on the pan at the end which carries the weights and these are removed by the dials to give a balancing weight of 200 g. This ensures that the beam is always loaded to the same extent. Errors due to deformation of the beam caused by varying loads are avoided in this way.

Balances differ in the way in which the final weight is obtained and displayed. A common system is for the total swing of the balance to correspond to a weight difference of 0.1000 g. A calibrated graticule numbered from 0 to 1000 is attached to the end of a lever fixed to the centre of the beam. Using a simple optical system, the image of a small part of the scale is focused on a ground glass screen carrying a single line. The zero of the scale is brought into line with this line. Unit and first decimal place fractional weights are arranged on the beam to balance the object to the nearest 0.1 g. The last three decimal places are read directly from the illuminated scale.

Developments in electronics are currently revolutionising the weighing operation. Fully electronic balances are now available for covering the ranges of weights provided by beam balances. These new balances have no weights, no beam or knife edges, no weight set and no moving parts. Thus, errors associated with these traditional features are eliminated. They may be used to weigh, count, or read directly in percentage or units other than grams. These and many other functions use programmes which are stored in a microprocessor either within the balance or supplied as an accessory.

The new balances are very much less dependent on the laboratory environment and the experience of the analyst. It is certain that they will replace virtually all traditional beam balances within the next few years.

4

PRELIMINARY TREATMENT OF THE SAMPLE

Once a representative sample has been taken, a quantitative analysis proceeds through a number of stages. These vary according to the nature of the material and the desired analysis. Sometimes one or two of the stages may be omitted.

Measure the analytical sample by weight or by volume.
Dissolve in a suitable solvent.
Separate the desired component from interfering substances.
Measure the desired component.
Calculate the result by relating the final measurement to the sample weight or volume.

In the case of single chemicals it may be possible to dissolve the analytical sample in water or, say, dilute acid. However, some substances are difficult to dissolve and require more elaborate treatments. With more complex materials it may be necessary to use separation techniques such as precipitation, solvent extraction or chromatography to isolate the desired component in a form suitable for measurement.

It is impossible to describe all preliminary treatments since the variety of materials and desired analyses is virtually limitless. Here are a few examples.

4.1 TREATMENT OF A SPARINGLY SOLUBLE SUBSTANCE

Hydrochloric acid, nitric acid or 'aqua regia' (1 volume of nitric acid + 3 volumes of hydrochloric acid) dissolve many inorganic substances which are insoluble in water.

Perchloric acid will dissolve some inorganic substances which fail to respond to the more familiar acids. It is a powerful oxidising agent and is specially useful for dissolving alloy steels in which chromium (VI) is quantitatively produced and may be determined very easily.

Perchloric acid must *not* be used for treating organic substances except under carefully proven and specified experimental conditions. Neither must it be allowed to come into contact with reducing agents such as inorganic salts of low oxidation state, for example, tin (II) compounds.

Fusion in a platinum crucible with potassium pyrosulphate is effective for certain ores such as crude titanium oxide. This is equivalent to a powerful acid attack at high temperature. The melt is allowed to solidify and is then dissolved in water or dilute acid.

Sometimes a fusion with sodium or potassium carbonate in a platinum vessel is used, especially when one of the sample components forms an insoluble carbonate. The melt is allowed to cool and leached out of the vessel with water. Insoluble carbonates of metals such as barium and zinc may be filtered, leaving a solution containing the soluble sodium salts of the anions formed by elements such as silicon and aluminium.

The majority of organic substances are insoluble in water, but are often soluble in organic solvents such as ethanol. These may be chosen to suit the intended procedure.

4.2 OXIDATION OF ORGANIC SUBSTANCES

The oxygen flask technique is often used for decomposing organic substances especially when halogens or sulphur are to be determined. A small sample of 20–30 mg weight is wrapped in a square of paper, or if a liquid, is placed in a gelatine capsule. It is placed in a platinum basket suspended from a stopper and is ignited in a closed flask filled with oxygen. The flask contains a small volume of dilute alkali. The combustion products, carbon dioxide, water, possibly oxides of nitrogen, hydrogen halides and oxides of sulphur are absorbed. A few drops of hydrogen peroxide may be added to ensure the oxidation of sulphur compounds to sulphate. The latter, and halide if present, is then determined by an inorganic titration.

An alternative approach suitable for determining inorganic constituents other than halogens or sulphur is to 'wet oxidise' the sample with hot concentrated sulphuric acid. A long-neck, round-bottom (Kjeldahl) flask is used for this purpose. Carbon and hydrogen are converted to carbon dioxide and water and the inorganic constituents remain as sulphates. These may then be determined by standard inorganic procedures. Nitrogen if present is converted to ammonium sulphate and may be determined in this form. The solution is made alkaline with sodium hydroxide and boiled. The resultant ammonia gas is passed into an excess of standard acid. The excess is measured by a back titration with standard alkali and the nitrogen content is calculated with reference to the sample weight.

4.3 THE SEPARATION OF METAL IONS WITH ORGANIC REAGENTS

It is often necessary to separate interfering metals prior to analysis. It is also useful to transfer the metal from aqueous to non-aqueous solution. This is especially valuable in some atomic absorption spectroscopy methods, when the sensitivity of the method can be considerably increased.

Many metal ions react with a certain type of organic compound to form an *organo-metallic complex*. For this to be possible, the organic compound must:

Have an acidic hydrogen atom replaceable by a metal.

Have within its molecule an atom such as nitrogen or oxygen which has an unbonded electron pair.

Be capable of forming at least one ring of atoms, five or six in number and including the metal and the atom with the unbonded pair of electrons.

The mechanism may be understood by considering the simplest example, glycine or aminoacetic acid. The molecular structure may be written like this:

$$\underset{H_2N\,:}{\overset{\displaystyle CH_2-C}{\diagup}}\overset{\displaystyle O}{\underset{\underset{H}{\diagup}O}{\diagdown}}$$

The nitrogen atom has five electrons in its outer shell. Only three of these are being used. The unbonded pair is available for further combination.

In alkaline conditions, the glycine molecule loses its acidic hydrogen to form a glycinate anion.

$$\underset{H_2N\,:}{\overset{\displaystyle CH_2-C}{\diagup}}\overset{\displaystyle O}{\underset{O^{\ominus}}{\diagdown}}$$

Two glycinate anions combine with one copper (II) cation to form a copper (II) glycinate complex:

$$2\,NH_2.CH_2.COO^- + Cu^{2+} \rightarrow Cu(NH_2.CH_2.COO)_2$$

The copper (II) cation has:

Replaced a hydrogen ion in each of two molecules of glycine.
Linked with each of the nitrogen atoms by utilising the unbonded electron pair in each.

This type of substance is called a *chelate* compound and the organic compound is called a *chelating agent*.

Many such compounds are selective with regard to the metals with which they will combine or may be made so by carefully selecting the acidic or basic conditions (pH). Chelate compounds are generally more soluble in organic solvents than in water, and may be separated from other metals which do not form a complex in the chosen conditions.

Here are some examples:

8-hydroxyquinoline (oxine) forms complex with aluminium ions.
Ammonium pyrrolidine dithiocarbamate forms complexes with copper, iron and lead ions.
N-nitroso-phenylhydroxylamine ammonium salt (Cupferron) forms complexes with copper and iron ions.
Dimethylglyoxime forms complexes with nickel and palladium ions.

5

PHYSICAL PROPERTIES

The most useful of these are density, melting point, refractive index and where applicable, optical rotation. In each case, a perfectly pure specimen of a single chemical will yield definite and unchanging values if the determinations are carried out in standard conditions. A deviation will therefore indicate the presence of one or more impurities. The values are also useful for defining the properties of mixtures within specified limits.

5.1 DENSITY

This value is defined as the weight per unit volume at a certain temperature, for example the weight in grams per cm^3 at $20°C$. For liquids, the value is determined by accurately weighing a known volume at the appropriate temperature in a specially designed narrow neck graduated bottle. Alternatively a 'density meter' may be used. This depends on the fact that the natural frequency of a hollow oscillator is changed when filled with a liquid. The instrument is calibrated against air and water and gives a direct read-out of density.

5.2 MELTING POINT

The value for a pure substance is usually lowered by the presence of impurities. Values for many pure substances may be found in the common reference books such as 'The Handbook of Chemistry and Physics' and 'The Merck Index'.

The classical method for determining this value is to heat a little of the finely powdered substance in a glass capillary tube sealed at one end. The tube is attached to a thermometer and both are heated at a controlled rate in a liquid bath. Several heating liquids are suitable, including liquid paraffin. The temperature is noted when the substance begins to soften, when a meniscus first forms and when the whole of the substance has become liquid.

In addition to lowering the melting point, impurities also extend the range of temperature over which the melting process takes place. Thus a narrow melting range which includes the value for the pure substance is a good indication of high purity. Consequently, a wide and low melting range generally indicates low purity.

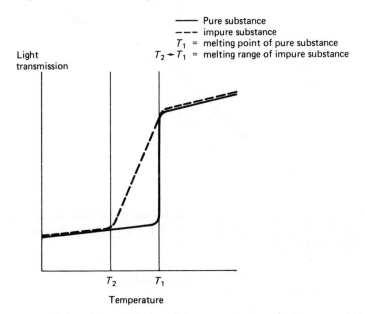

Figure 1 *The melting of pure and impure substances using an optical detector*

There are several pieces of equipment which enable the melting point or melting range to be determined semi-automatically, thus avoiding differences in technique which could produce different results. In one of these, a narrow beam of light is directed on to a capillary tube containing the substance. A photoelectric sensor is situated behind the tube. The tube is heated. When the substance melts, the light beam passes through the liquid and activates the photoelectric sensor. The response signal is processed and stops the heating. The melting point is read from a temperature read-out meter. Arrangements may also be made for recording the melting range on a moving chart recorder (figure 1).

5.3 REFRACTIVE INDEX (figure 2)

When a beam of light passes from one medium to another, it undergoes *refraction*, i.e. a change of direction. This is due to a change in velocity.

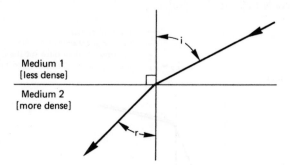

Figure 2   *Refractive index*

For example, if a beam of light travels from air or a vacuum to a more dense medium, for example, water, then if i = angle of incidence, r = angle of refraction,

$$\frac{\text{Sin } i}{\text{Sin } r} = \text{refractive index}$$

The value, which is always greater than 1, is the ratio of the velocities of light in the two media. It varies with temperature and with the wavelength of light. If either of the media is a gas, then the pressure will also affect the result.

The *absolute refractive index* of a substance is the value of $\sin i/\sin r$ when Medium 1 is a vacuum.

The absolute refractive index of air is 1.000294 and is so close to unity that it is commonly used as a more convenient standard than a vacuum.

The refractive index value varies inversely with the wavelength of the light and is greater at the violet (short wavelength) end of the spectrum than at the red (long wavelength) end. In practice the D spectral line of sodium, produced by a sodium vapour lamp, is used. The refractive index value so determined approximates to the mean value for the multiple wavelengths of white light. Values are generally measured at 20°C.

The symbol is n_D^{20} which signifies the refractive index value relative to air, measured at 20°C, with sodium D light of wavelength 589.3 nm (589.3 $\times$ 10^{-9} m).

The value is determined with a *refractometer* which is usually calibrated against 1-bromonaphthalene.

5.4 OPTICAL ROTATION

White light produced from a tungsten filament lamp consists of a mixture of different wavelengths distributed evenly throughout the visible range of the spectrum. It is *polychromatic*. A selected narrow waveband of light containing only a very few wavelengths is called *monochromatic*.

The wavelength of a beam of light may be visualised symbolically by comparing it with a rope. One end is attached to a fixture and the free end is held in the hand. If the free end is waved up and down in a vertical plane, waves will pass along the rope in this plane, the wavelength depending on the frequency with which the rope is waved. If the wavelength is maintained constant, the rope represents *plane polarised monochromatic light*.

A beam of light, as normally produced, whether monochromatic or polychromatic, gives rise to planes of wave motion extending through the full circle of 360°. Certain crystals, by virtue of their lattice structures and dimensions, have the ability to convert a beam of normal light into *polarised light* in which the wave motion is in one plane. In the rope analogy, the crystal behaves like a vertical slit superimposed between the free and fixed ends. If the rope is waved at random in all directions, only the vertical waves will pass through the slit.

Now some substances, most of them organic, possess a special optical property when dissolved in a suitable solvent. They rotate the plane of a beam of plane polarised monochromatic light.

The angle of rotation is related to:

The concentration of the solution.
The path length of the solution through which the light travels.
The wavelength of the light.
The chemical configuration of the substance.

The behaviour is called *optical activity* and such substances are said to be *optically active*.

Optically active substances have one property in common. They have *asymmetric molecules* which may exist in two different forms, one being the mirror image of the other. The criterion of asymmetry is that it shall not be possible to superimpose one molecular structure upon its mirror image so that all the atoms coincide with each other in their relative positions in space.

One of the commonest causes of asymmetry is an asymmetrically substituted carbon atom (figure 3). The four groups attached to a carbon atom are directed towards the corners of a tetrahedron. If the groups are all different it may be seen that the mirror images cannot be exactly superimposed.

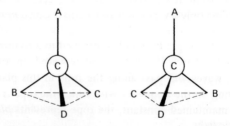

Figure 3 *Asymmetrically substituted carbon atoms*

The two different forms of such a substance are called *stereoisomers* and the behaviour is called *stereoisomerism*. They are designated (+) or (−) according to the direction in which the plane polarised light is rotated, i.e. clockwise or anticlockwise. The extent of the rotation is exactly the same for the two forms, but in opposite directions.

A mixture of equal proportions of the (+) and (−) forms gives a *racemic mixture*, the (±) form. This does not show optical activity because the activities of the (+) and (−) forms cancel each other out.

For a stated set of experimental conditions, the angle of rotation is a characteristic for a given optically active substance. The value is normally expressed as the Specific Optical Rotation at 20°C using the Sodium D line. The symbol is $[\alpha]_D^{20}$.

If α = observed angle of rotation
                                      c = concentration in g/100 cm^3
                                      l = path length in decimetres

then $[\alpha]_D^{20} = 100\alpha/lc$ = angle of rotation which would be caused by passing a beam of plane polarised sodium D light through one decimetre of a solution containing 100 g of solute per 100 cm^3 of solution. The measurement is made with a *polarimeter* (see Plate 2 and figure 4).

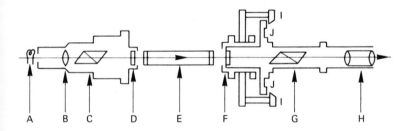

Figure 4 *The polarimeter*

Light from a sodium vapour lamp (A) is passed through a collimator lens (B) to produce a parallel beam. The beam is passed through the polarising Nicol prisms (C).

The polarised beam is passed via the glass window (D) through the sample tube (E), via a window (F) through a second Nicol prism (G) and finally through a telescope (H) through which it is viewed.

The assembly containing F, G and H may be rotated and its position may be accurately noted by means of the fixed viewers (II) in relation to a circular scale calibrated in degrees shown in section (JJ).

The prisms are first arranged so that in the absence of a sample, no light passes through to the telescope.

The sample solution is interposed between the prisms, the plane of polarised light is rotated and some light passes through the second prism.

The second prism is rotated until once again no light passes.

The angle of rotation will equal the angle by which the plane polarised light has been rotated by the solution.

6

CLASSICAL ANALYSIS

This term is used to describe those procedures which depend on the measurement of chemical reactions with simple equipment including the balance, graduated flasks and graduated measuring devices such as the pipette and the burette (see Plates 3 and 4). It also includes visual calorimetry in which colours are matched by eye rather than instrumentally. In spite of the enormous development in instrumentation in recent years, classical analysis still plays an important part. In particular, classical methods often provide the only means of achieving the desired accuracy and precision for major component analysis.

6.1 TITRATIONS – BASIC PRINCIPLES – THE MOLE CONCEPT

An accurately known weight or volume of sample is dissolved in a suitable solvent. A solution of a reacting substance is run into the sample solution from a burette until the reaction is just complete when the volume is measured. The completion of the reaction is marked by the change in colour of an indicator. Alternatively, the end-point may be detected instrumentally.

The solution of reacting substance is of known concentration, and is called the *titrant* or *standard solution*. From a knowledge of the chemical reaction, the amount of reacting substance and the sample weight or volume, we may calculate the analytical result. This may be expressed for example as a weight percentage or as a weight per unit volume.

In *titrimetric analysis*, we usually express the concentration of the titrant in terms of *molarity* or *normality* because this greatly simplifies the calculations. This requires an understanding of the *mole concept*.

1 mole is defined as the amount of substance which contains the same number of species as there are atoms in 12 g of the carbon isotope ^{12}C.

For all elements, 1 mole is the average atomic mass expressed in grams. For example,

> 39.10 g of potassium contains 1 mole of K atoms.
> 55.85 g of iron contains 1 mole of Fe atoms.

For any element which exists as molecules or for any compound 1 mole of molecules is the *molecular mass* expressed in grams. For example,

> 32.00 g of oxygen contains
> 1 mole of O_2 molecules, but
> 2 moles of O atoms.

> 58.44 g of sodium chloride contains
> 1 mole of NaCl 'molecules'.

It also contains   1 mole of Na^+ ions
and   1 mole of Cl^- ions.

A molar solution contains 1 mole of the atoms or molecules of the stated substance dissolved in 1 dm^3 of solution. It may be designated 'M' or '1 mol/dm^3'. For example, a molar (M) solution of hydrochloric acid contains 1 mole of HCl 'molecules' per dm^3 = 36.46 g HCl/dm^3. It also contains 1 mole of H^+ ions and 1 mole of Cl^- ions per dm^3.

A molar (M) solution of sulphuric acid contains 1 mole of H_2SO_4 'molecules' per dm^3 = 98.08 g H_2SO_4/dm^3. It also contains 2 moles of H^+ ions and 1 mole of SO_4^{2-} ions per dm^3.

A molar (M) solution of sodium hydroxide contains 1 mole of NaOH 'molecules' per dm^3 = 40.00 g $NaOH/dm^3$. It also contains 1 mole of Na^+ ions and 1 mole of OH^- ions per dm^3.

Notice that one 'molecule' of H_2SO_4 provides two H^+ ions. Hence 1 mole of H_2SO_4 contains 2 moles of H^+ ions. This means that a molar (M) solution of sulphuric acid is twice as concentrated an acid as molar (M) hydrochloric acid.

In industrial analysis, it is often convenient to express solution strengths in terms of *equivalent concentrations*, or *normalities* since this further simplifies the calculations. At one time, this was universal practice, but the idea was abandoned by the schools and universities many years ago. Nevertheless, the concept, which is no more than a special application of the mole concept, is given here.

A normal (N) solution contains 1 mole of the reacting species per dm^3.

The quantity of substance (the equivalent weight) required to prepare 1 dm^3 of N solution may vary. It is that mass of substance, expressed in grams, which provides 1 mole of reacting species in the specific reaction. It is calculated by dividing the molecular mass by the number of reacting species provided by one molecule of the substance.

Example 1. Titration of sodium hydroxide with hydrochloric and sulphuric acids.

The reacting species are the OH^- ion and the H^+ ion.

Since $NaOH \rightarrow OH^-$,   then equivalent weight $= \dfrac{\text{molecular mass}}{1}$ g

$$= 40.00 \text{ g}$$

Since $HCl \rightarrow H^+$,   then equivalent weight $= \dfrac{\text{molecular mass}}{1}$ g

$$= 36.46 \text{ g}$$

Since $H_2SO_4 \rightarrow 2H^+$,   then equivalent weight $= \dfrac{\text{molecular mass}}{2}$ g

$$= \dfrac{98.08}{2} = 49.04 \text{ g}$$

For all neutralisation reactions in aqueous solution, the equivalent weight of the acid or base may be defined as:

$$\dfrac{\text{Molecular mass in g}}{\substack{\text{Number of } H^+ \text{ or } OH^- \text{ provided or accepted by one molecule} \\ \text{or ion in the particular reaction}}}$$

Example 2. Reduction/oxidation titration of iron (II) sulphate with potassium permanganate.

The reacting species is the electron and the equivalent weight becomes:

$$\dfrac{\text{Molecular mass in g}}{\substack{\text{Number of electrons provided or accepted by one molecule} \\ \text{or ion in the particular reaction}}}$$

Since $Fe^{2+} \rightarrow Fe^{3+} + 1$ electron, then equivalent weight of

$$FeSO_4\,7H_2O = \frac{\text{molecular mass}}{1}\ g$$

$$= 278.01\ g$$

Since $Mn^{7+} + 5$ electrons $\rightarrow Mn^{2+}$, then equivalent weight of

$$KMnO_4 = \frac{\text{molecular mass}}{5}\ g$$

$$= \frac{158.03}{5} = 31.61\ g$$

The number of electrons provided or accepted is the change in oxidation number of the key element and so the equivalent weight of an oxidising or reducing agent becomes:

$$\frac{\text{Molecular mass in g}}{\text{Change in oxidation number of key element}}$$

An explanation of oxidation number is given in section 6.4.

The advantage of adopting the normality/equivalent weight concept is that 1 cm³ of normal (N) solution of Solution A is equivalent to 1 cm³ of normal (N) solution B. This greatly simplifies calculations in an industrial analytical laboratory. The disadvantage is that some substances may provide or require different numbers of reacting species per molecule in different circumstances. This means that the equivalent weight may vary. In this situation, the strength of a solution should always be calculated and expressed in terms of its molarity (M).

6.2 SIMPLE ACID–BASE TITRATIONS – THE pH SCALE

We have seen in section 6.1 that neutralisation reactions in aqueous solution depend on the reaction between H^+ ions provided by the acid and OH^- ions provided by the base.

Water is itself a weak electrolyte and dissociates slightly into hydroxonium and hydroxyl ions:

$$2H_2O \rightleftharpoons H_3O^+ + OH^-$$

For simplicity, the hydroxonium ion may be regarded as a hydrated hydrogen ion and the dissociation may be expressed:

$$H_2O \rightleftharpoons H^+ + OH^-$$

At 20°C, the concentrations in mol/dm³ of hydrogen and hydroxyl ions are each 10^{-7}, i.e.

$$[H^+] = [OH^-] \text{ (square brackets designate 'concentration')}$$

The *ionic product* $[H^+]$ $[OH^-]$ = $10^{-7} \times 10^{-7}$ = 10^{-14} mol²/dm⁶ at 20°C. This is a constant at this temperature. It is given the symbol K_w.

In a neutral solution, $[H^+] = [OH^-] = 10^{-7}$ mol/dm³
In an acidic solution, $[H^+] > 10^{-7}$ mol/dm³
In a basic solution, $[H^+] < 10^{-7}$ mol/dm³

It is cumbersome to express hydrogen ion concentrations in this way.
Instead, the pH is used (power of the hydrogen ion concentration). It is the power (log) to base 10 with the sign changed:

$$pH = - \log_{10} [H^+]$$

In a neutral solution,   pH = 7
In an acidic solution,   pH = 0 → 7
In a basic solution,   pH = 7 → 14

At other temperatures, the ionic product of water has different values. Hence the pH scale and the neutral pH value will vary from the values we normally work with at 20°C.

For strong acids and strong bases in which we may assume virtually one hundred per cent dissociation to provide H^+ and OH^- ions, the pH values are as follows:

		pH
Hydrogen ion 1	mol per dm³ (M)	0
Hydrogen ion 0.1	mol per dm³ (0.1 M)	1
Hydrogen ion 0.01	mol per dm³ (0.01 M)	2
Hydroxyl ion 1	mol per dm³ (M)	14
Hydroxyl ion 0.1	mol per dm³ (0.1 M)	13
Hydroxyl ion 0.01	mol per dm³ (0.01 M)	12

In an acid-base titration in aqueous solution we either titrate hydrogen ion with standard base (hydroxyl ion) or hydroxyl ion with standard acid (hydrogen ion).

If we plot the change in pH with added standard solution, we get a titration curve (figure 5). The shape of the curve depends upon the acid and base. The point at which the change in pH is a maximum is the inflection point. In most cases it marks the end-point of the titration or the equivalence point.

The end-point of a titration may be detected either by following the change in pH with a pH meter or by using a visual indicator.

An acid–base indicator is a substance with a colour which depends on the pH of the solution. It is itself a weak acid or weak base and the change in colour is due to a change in molecular structure when a

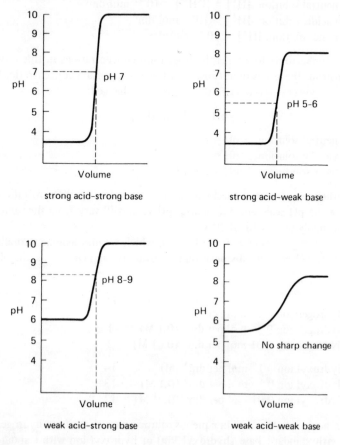

Figure 5 *Titration curves of acids and bases*

hydrogen ion is gained or lost. Different indicators change colour at different pH values and the correct one must be chosen to match the equivalence pH of the titration.

For example, phenolphthalein changes its structure at pH 9.7. It is colourless below pH 9.7 (In_A) and violet-pink above pH 9.7 (In_B):

$$In_A \rightleftharpoons In_B + H^+$$
$$\text{colourless} \quad \text{reddish-violet}$$

Table 1 Some acid–base indicators

Name	pH at mid-colour change	Colour change acid–base
Cresol red	1.0	red–yellow
Thymol blue	2.0	red–yellow
Methyl orange	3.8	red–yellow
Bromocresol green	4.6	yellow–blue
Methyl red	5.3	red–yellow
Bromocresol purple	6.0	yellow–purple
Bromomethyl blue	6.8	yellow–blue
Phenol red	7.5	yellow–red
Phenolphthalein	9.2	colourless–reddish-violet
Thymolphthalein	10.0	colourless–blue

6.3 NON-AQUEOUS TITRATIONS

An acid is a proton donor and a base is a proton acceptor. Water itself acts as a base for the reception of protons from a strong acid such as HCl which may then be titrated with, for example, sodium hydroxide:

$$HCl + H_2O \rightleftharpoons H_3O^+ + Cl^-$$
$$\text{acid} \quad \text{base} \quad \text{acid} \quad \text{base}$$

The neutralisation reaction may be represented:

$$H_3O^+ + OH^- \rightarrow 2 H_2O$$
$$\text{acid} \quad \text{base}$$

The hydroxonium and hydroxyl ions are special examples of proton donor (acid) and proton acceptor (base) respectively.

Many other liquids may be used as solvents for acid–base titrations and use may be made of their acidic or basic characteristics which may differ markedly from that of water.

The most useful solvents are either strongly basic (proton-accepting, protophilic) or they resemble water in that they may accept or donate protons according to the nature of the solute (amphiprotic).

1:2-diamino Ethane (Protophilic)

A weakly acidic substance, often sparingly soluble in water, will dissolve and dissociate in the solvent to give a highly acidic solvated proton which may be titrated with a strong base such as methanolic KOH. For example, a high molecular weight carboxylic acid:

$$2R.COOH + NH_2.CH_2.CH_2.NH_2 \rightarrow 2R.COO^- + NH_3^+.CH_2.CH_2.NH_3^+$$
   acid   base   base   acid

$$NH_3^+.CH_2.CH_2.NH_3^+ + 2\ OH^- \rightarrow NH_2.CH_2.CH_2.NH_2 + H_2O$$
     acid   base   base   acid

Note that on neutralisation, the undissociated basic solvent is regenerated. The solvated proton is analogous to the hydroxonium ion in aqueous solutions when, on neutralisation, the undissociated water molecule is regenerated.

Acetic Acid (Amphiprotic)

By analogy with the 'auto ionisation' of water (see section 6.2), this solvent behaves like this:

$$2\ CH_3.COOH \rightleftharpoons CH_3.COOH_2^+ + CH_3COO^-$$

The cation and anion are analogous to the hydroxonium and hydroxyl ions in the water system. However, it is a stronger acid (proton donor) than water.

Consequently, many weak bases, too weak and sparingly soluble in water to be titrated in aqueous solutions, will dissolve and dissociate in acetic acid:

$$RNH_2 + CH_3COOH \rightarrow RNH_3^+ + CH_3.COO^-$$
   base   acid   acid   base

On the other hand, some very strong acids such as perchloric acid behave as though acetic acid were a base.

$$HClO_4 + CH_3.COOH \rightarrow ClO_4^- + CH_3.COOH_2^+$$
$$\text{acid} \qquad \text{base} \qquad \text{base} \qquad \text{acid}$$

It is thus possible to dissolve a weak base such as aniline in acetic acid and titrate it with a standard solution of perchloric acid in acetic acid. The neutralisation reaction is:

$$CH_3.COO^- + CH_3.COOH_2^+ \rightarrow 2\ CH_3.COOH$$

Notice that this is the 'auto ionisation' reaction in reverse, just as the neutralisation of hydroxonium ion by hydroxyl ion is 'auto ionisation' in reverse in aqueous solutions.

Several visual indicators are very effective, and include crystal violet, methyl violet and oracet blue B. The mechanism of the colour change is as described in section 6.2. The indicator changes its structure when it gains or loses a proton in passing from acid to basic solution conditions. Each structure has its characteristic light absorbance and this in turn produces a visual colour change.

6.4 REDOX TITRATIONS – OXIDATION NUMBER

In every reduction–oxidation (redox) reaction, the reductant donates electrons to the oxidant. The reductant is thereby oxidised and the oxidant is thereby reduced. The number of electrons involved in relation to the numbers of reacting species depends on the change in oxidation number suffered by the reductant and the oxidant.

Oxidation = loss of electrons and increase in oxidation number.

Reduction = gain of electrons and decrease in oxidation number.

Many useful titrations are based on reduction or oxidation, employing a suitable standard solution.

Oxidation Number (O.N.)

This number represents the electric charge, (which may be positive or negative) carried by a species, assuming certain rules in assigning electrons within compounds. It does not necessarily represent an exact state, but is rather a practical aid to working out the stoichiometry of redox reactions. The rules are as follows:

1. The overall O.N. of a neutral species is 0.
2. The overall O.N. of a charged species equals the charge,

e.g. O.N. of NaCl = 0   O.N. Na$^+$ = +1
                           O.N. Cl$^-$ = −1
e.g. O.N. of NO$_3^-$ = −1   O.N. N = +5 and O.N. O = −2.

3. The O.N. is not changed on the formation of a complex ion, e.g. O.N. of cobalt = +3 in $[Co(NH_3)_6]^{3+}$ and $[Co(CN)_6]^{3-}$.
4. The O.N. of combined oxygen is −2 except in peroxides when it is −1.
5. The O.N. of combined hydrogen is +1 except in the metallic hydrides when it is −1.

Let us see how the concept works out with some well known substances.

Potassium permanganate. What is the O.N. of the manganese atom in $KMnO_4$?

$$
\begin{aligned}
\text{O.N. of } K^+ &= +1 \\
\text{O.N. of } MnO_4^- &= -1 \\
\text{O.N. of each O atom} &= -2 \\
\text{O.N. of four O atoms} &= -8
\end{aligned}
$$

The sum of the O.N. values of the Mn atom and the four O atoms must equal −1, the value for MnO_4^- ion.

$$
\begin{aligned}
\text{Let O.N. Mn atom} &= x \\
\text{Then } x + (-8) &= -1 \\
x - 8 &= -1 \\
\therefore \quad x &= -1 + 8 = +7
\end{aligned}
$$

Potassium dichromate. What is the O.N. of the chromium atom in $K_2Cr_2O_7$?

$$
\begin{aligned}
\text{O.N. of } K^+ &= +1 \\
\text{O.N. of two } K^+ &= +2 \\
\text{O.N. of } Cr_2O_7^{2-} &= -2 \\
\text{O.N. of each O atom} &= -2 \\
\text{O.N. of seven O atoms} &= -14
\end{aligned}
$$

The sum of the O.N. values of the two Cr atoms and the seven O atoms must equal −2, the value of the $Cr_2O_7^{2-}$ ion.

$$
\begin{aligned}
\text{Let} \qquad \text{O.N. Cr atom} &= x \\
\text{Then} \qquad 2x + (-14) &= -2 \\
2x - 14 &= -2 \\
2x &= +12 \\
\therefore \qquad x &= +6
\end{aligned}
$$

Let us look at a typical redox titration, say of iron (II) with standard potassium permanganate solution. We write down the 'half reactions' for the iron (II) (reductant) and the MnO_4^- ion (oxidant).

$$Fe^{2+} \rightarrow Fe^{3+} + e^{1-}$$

Change in O.N. of iron $= (3+) - (2+) = +1$ corresponding with a *loss* of one electron. The titration is carried out in dilute sulphuric acid and the permanganate is reduced to Mn (II) sulphate (we need to know this)

$$MnO_4^- \rightarrow Mn^{2+}$$

The O.N. of the manganese atom in MnO_4^- is +7 (as we have seen), and the O.N. of the manganese atom in Mn^{2+} is +2. The change in O.N. is therefore -5, involving a gain of five electrons. So we may write

$$MnO_4^- + 5e^- \rightarrow Mn^{2+}$$

(or more simply: $Mn^{7+} + 5e^- \rightarrow Mn^{2+}$)
If we wish, we may complete the rest of the 'half reaction'.

Balance the oxygen atoms on the left by adding water molecules on the right. Then balance the hydrogen atoms introduced by the water molecules by adding hydrogen ions on the left.

$$8H^+ + MnO_4^- + 5e^- \rightarrow Mn^{2+} + 4H_2O$$

This 'balances' with respect to electrical charges as well as mass.

The complete reaction may then be written by adding the two half reactions. First of all, multiply the iron half reaction by five so that the electrons cancel out.

$$5Fe^{2+} + MnO_4^- + 8H^+ \rightarrow 5Fe^{3+} + Mn^{2+} + 4H_2O$$

It is not actually necessary to work out the complete reaction to calculate the required concentration of a redox standard solution if we apply the normality/equivalent weight concept. It is necessary only to know the initial and final oxidation numbers of the species undergoing oxidation or reduction as described at the end of section 6.1.

Equivalent weight of an Fe (II) solution $= \dfrac{\text{Molecular Mass}}{1}$ g

Equivalent weight of a $KMnO_4$ solution $= \dfrac{\text{Molecular Mass}}{5}$ g

So a decinormal (0.1 N) solution of, say, iron (II) sulphate is also a decimolar (0.1 M) solution, but a decinormal (0.1 N) solution of potassium permanganate is a decimolar/5 solution. Equal volumes of each solution are equivalent, i.e. they react exactly with each other.

Again it must be stressed that although this concept is very useful, the change in oxidation number of a given atom is not always the same. For this reason, if a solution, say of potassium permanganate, is being used in different circumstances, its strength must always be expressed as an unambiguous molarity.

6.5 TITRATIONS WITH EDTA

EDTA is the abbreviated name for ethylenediaminetetra-acetic acid.

$$
\begin{array}{cc}
\text{HO.CO.CH}_2 & \text{CH}_2.\text{CO.OH} \\
\diagdown & \diagup \\
\text{N} - \text{CH}_2 - \text{CH}_2 - \text{N} \\
\diagup & \diagdown \\
\text{HO.CO.CH}_2 & \text{CH}_2.\text{CO.OH}
\end{array}
$$

The anion of this substance is able to form stable, water-soluble complexes with a large number of metal ions (figure 6). The principle has been used to devise a wide range of methods for the direct titration of metals. In practice, the disodium salt is used because it is readily soluble in water and the solution is indefinitely stable. It exists as a dihydrate, which is important to remember when preparing standard solutions.

$$
\begin{array}{cc}
\text{HO.CO.CH}_2 & \overset{\ominus\oplus}{\text{CH}_2.\text{CO.O Na}} \\
\diagdown & \diagup \\
\text{N} - \text{CH}_2 - \text{CH}_2 - \text{N} \\
\overset{\oplus\ominus}{\underset{\text{Na O.CO.CH}_2}{\diagup}} & \diagdown \\
 & \text{CH}_2.\text{CO.OH}
\end{array}
$$

We may represent the salt as Na_2H_2Y giving rise to the reacting ion H_2Y^{2-}. This reacts in the following way with ions of different charges:

$$M^{2+} + H_2Y^{2-} \rightleftharpoons MY^{2-} + 2H^+$$

$$M^{3+} + H_2Y^{2-} \rightleftharpoons MY^- + 2H^+$$

$$M^{4+} + H_2Y^{2-} \rightleftharpoons MY + 2H^+$$

Notice that in each case, one metal ion reacts with one EDTA ion, which makes the calculations easy. Also, two hydrogen ions are liberated.

M²⁺ ion: stable in neutral or alkaline buffer solutions

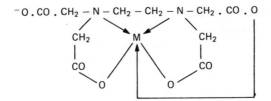

M³⁺ ion: stable in fairly acidic, neutral or alkaline buffer solutions

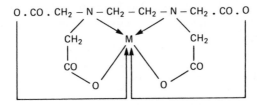

M⁴⁺ ion: stable in strongly acid, neutral or alkaline buffer solutions

Figure 6 *The structures of the EDTA complexes*

These are neutralised by adding a buffer solution to prevent the back reaction from taking place. The metal ion displaces the hydrogen ions at opposite ends of the molecule and is co-ordinated by the unshared electron pair on each nitrogen atom. In addition, M^{3+} and M^{4+} ions form additional bonds with the unshared electron pairs on the terminal oxygen atoms. This partly accounts for the greater stability of the complexes of, say, aluminium and zirconium compared with those of calcium, magnesium and zinc. Advantage may be taken of this difference in behaviour. The reaction of M^{2+} ions may be suppressed by using a low pH buffer, enabling M^{3+} and M^{4+} ions to be selectively titrated.

The general equation for complex formation may be written:

$$M^{n+} + H_2Y^{2-} \rightleftharpoons MY^{(n-4)} + 2H^{1+}$$

If we remove the hydrogen ions (which take no direct part in complex formation), the equation becomes:

$$M^{n+} + Y^{4-} \rightleftharpoons MY^{(n-4)}$$

From the Law of Mass Action:

$$\frac{[MY^{(n-4)}]}{[M^{n+}][Y^{4-}]} = \text{Constant} = K \text{ the Stability Constant}$$

The square brackets designate the concentrations of the reactants and the product in mol/dm^3.

The value of K gives a measure of the stability of a particular metal–EDTA complex. This enables us to choose the most appropriate buffer solution and indicator. A comparison of K values also enables us to decide on the best conditions for titrating one metal ion in the presence of another.

K values are very high, from around 10^{10} upwards. For this reason, it is more convenient in practice to use $\log_{10} K$ values. We speak of $\log_{10} K = 10.5$, 14.6, 18.7, etc. The higher the value of $\log_{10} K$, the more stable is the complex and the lower is the pH at which the metal ion may be titrated.

6.5.1 Indicators for EDTA Titrations

A suitable indicator must:

(a) Form a complex with the metal with a different colour from that of the indicator itself.
(b) Have a $\log_{10} K$ value several units less than the $\log_{10} K$ value of the metal–EDTA complex.

Let us consider the behaviour of an indicator which gives a *blue* solution and which forms a *red* complex with a given metal ion. Initially we have:

$$\text{metal ion} + \text{indicator} \rightarrow \text{Metal-indicator complex}$$
$$\text{(colourless)} \quad \text{(blue)} \quad\quad\quad\quad \text{(red)}$$

As we titrate with EDTA, it displaces the indicator from the metal-indicator complex because the EDTA complex is so much more stable.

EDTA   + metal–indicator complex → metal–EDTA complex + indicator
(colourless)   (red)   (colourless)   (blue)

The end-point occurs when the last trace of red is replaced by a clear blue.

In some cases, a better result may be obtained by adding an excess of EDTA and *back titrating* the excess with a standard metal ion solution. The latter must be chosen carefully. The $\log_{10} K$ value for the metal-EDTA complex must be several units less than the value for the metal–indicator complex, to avoid a breakdown of the latter.

6.5.2 Standardisation of EDTA Solutions

It is convenient to use solutions of 0.1 M or 0.01 M strengths, depending on the amount of metal ion it is wished to titrate. Ideally, the solution should be standardised against the same metal, using identical experimental conditions. This is essential if the highest accuracy is required.

In practice, it is common to use very pure zinc metal (99.999%) as the standard and to assume negligible differences when employing the factor so determined to calculate the results for other metals.

6.6 POTENTIOMETRIC TITRATIONS

It is sometimes not possible to use a visual indicator to detect the equivalence point of a titration; for example the test solution may be coloured in such a way as to mask the colour change. On the other hand there may be present more than one titratable component as in a mixture of different acids. In such cases the titration may be conducted by following the change in electrical potential of an *indicator electrode* immersed in the test solution.

If a metal is immersed in a solution of its own ions, a balance (equilibrium) is set up between the tendency for the metal atoms to ionise and pass into solution and the tendency for the metal ions in solution to give up their positive charges and deposit as metal atoms. This produces an electrical potential, the relative value of which is characteristic for each metal and its ion at a given concentration. The metal and its ion is called an *electrode*. For example, the silver electrode may be represented:

$$Ag^+ + e^- \rightleftharpoons Ag$$

Electrodes may be placed in order of *Standard Electrode Potential*. This is the electrical potential produced by immersing a metal in a solution containing effectively 1 mol/dm^3 of its own ions. The value is expressed relative to the Standard Hydrogen Electrode which is taken as zero and is represented by the reaction:

$$H^+ + e^- \rightleftharpoons \tfrac{1}{2}H_2$$

The actual potential may be represented at 20°C by the equation:

$$E = E_0 + \frac{0.0591}{n} \log_{10} [M^{n+}]$$

where E_0 = Standard Electrode Potential, n = number of electrons gained or lost per ion in the electrode reaction, $[M^{n+}]$ = actual concentration of the metal ion in mol/dm^3.

Instead of immersing a metal in a solution of its own ions, we may immerse a platinum wire in a solution of metal ions of different oxidation states, e.g. Fe^{2+} and Fe^{3+}. The system acquires an electrical potential depending on the tendency to move towards the higher or lower oxidation state. The iron system may be represented:

$$Fe^{3+} + e^- \rightleftharpoons Fe^{2+}$$

The actual potential may then be represented at 20°C by the simplified equation:

$$E = E_0 + \frac{0.0591}{n} \log_{10} \frac{[Fe^{3+}]}{[Fe^{2+}]}$$

The square brackets again signify mol/dm^3 concentrations of the two ions. n is the electron change between the higher and lower oxidation states. E_0 is called the *Standard Redox Potential* and, as the equation indicates, is the electrode potential when the two ions are present at equal concentrations. It is a special case of Standard Electrode Potential.

Another special case is the *glass electrode* which consists of a hollow bulb of thin glass filled with a solution of constant hydrogen ion concentration. When this is immersed in a solution containing hydrogen ions, an electrical potential is set up across the glass membrane dependent upon the difference in concentrations on each side of the glass.

At 20°C the simplified equation becomes:

$$E = E_0 + 0.0591 \log_{10} [H^+]$$

Since pH is defined as $-\log_{10} [H^+]$, this may be further simplified to:

$$E = E_0 - 0.0591 \text{ pH}$$

In each of these cases the actual electrode potential E is seen to be dependent on the ionic concentration or in the case of reduction/oxidation systems, a ratio of ionic concentrations. In a potentiometric titration (figure 7), this is altered by the addition of a standard titrant solution and the change is plotted against volume of titrant added. The graph so obtained is the *titration curve*.

The change in electrical potential of the indicator electrode is followed by connecting it to a *reference electrode* of constant potential to form a *cell*. The change in EMF of the cell thus reflects the change in potential of the indicator electrode. The cell is connected in opposition to a source of direct current. The moveable contact P is adjusted so that the cell output exactly balances the source, and no current is recorded by the galvanometer. The potential difference between A and B equals that of the source and so the cell potential may be calculated from the ratio AP/AB. In modern instruments the cell potential may be read directly in either millivolts or (for acid–base titrations) in pH units.

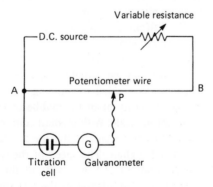

Figure 7 *Potentiometric titration arrangement*

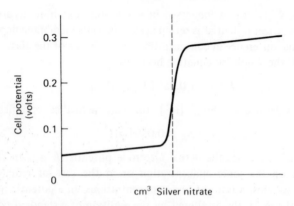

Figure 8 *Potentiometric titration of chloride with silver nitrate*

The indicator electrode is of bright silver wire and its potential depends upon the concentration of silver ions Ag^+. As standard silver nitrate solution is added, the Ag^+ ions are almost all removed from solution by the precipitation of insoluble silver chloride. The electrode potential therefore hardly changes. At the equivalence point, all the chloride ions will have been removed from solution and a further drop of standard silver nitrate solution produces a large relative change in the Ag^+ ion concentration. This produces a sharp change in cell (electrode) potential which is registered in the titration curve as an *inflection* (figure 8).

The equivalence point of the titration is taken to correspond with the middle of the inflection where the change in cell EMF per unit volume of standard silver nitrate solution is greatest.

It is often possible to titrate to a pre-selected millivolt or pH value, the flow of titrant being automatically stopped at this point. This has the advantage of eliminating personal judgement and variable experimental errors. Before this may be soundly implemented, the correct potential or pH must be determined by running a complete titration curve and determining the position of the middle of the inflection in relation to the millivolt or pH scale. With modern equipment, the whole titration may be accomplished automatically, the titration curve being drawn by a pen on a moving chart. The speed of the titration must be controlled so that it does not exceed the ability of the indicator electrode to adjust its equilibrium in relation to the test solution in which the ion concentration is constantly altered by the addition of standard titrant solution.

The Colour Plates

Plate 1   An analytical balance

Plate 2 A polarimeter

Plate 3 An analytical laboratory for classical analysis

Plate 4   A typical titration bench

Plate 5 A recording UV/visible spectrophotometer

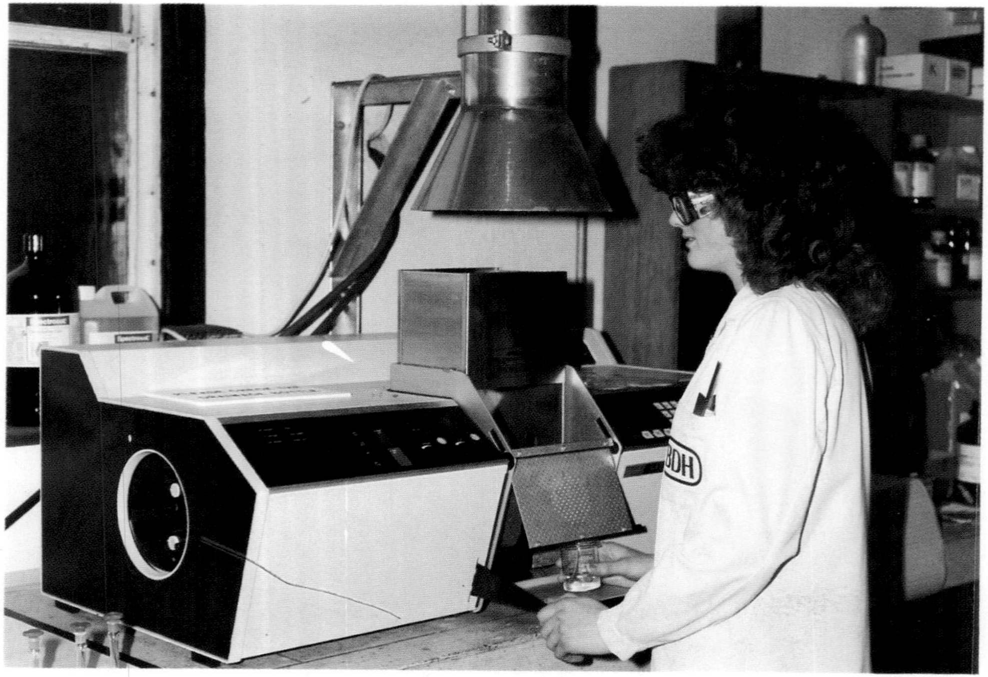

Plate 6 An atomic absorption spectrophotometer

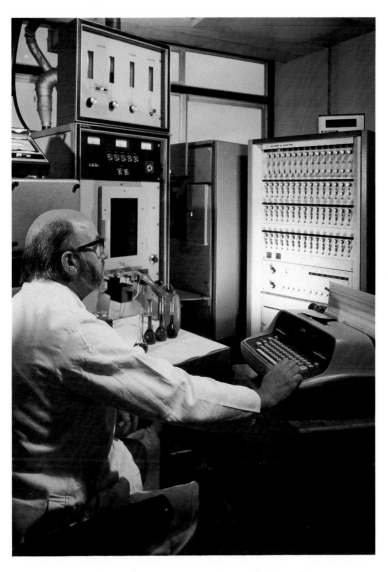

Plate 7 A plasma emission spectrograph

Plate 8 A gas chromatograph assembly

6.7 GRAVIMETRIC ANALYSIS

This is the oldest type of chemical analysis. It is still extremely useful in certain circumstances.

A known weight of sample is treated in such a way that a pure, virtually insoluble compound of the substance to be determined is precipitated from solution. The precipitate is filtered, washed free from soluble impurities, dried and weighed. Assuming a precise chemical composition of the precipitate, we can calculate the percentage of the substance in the sample.

Several important factors must be satisfied to ensure an accurate result.

(a) The precipitation must be complete.

The precipitate must be virtually insoluble. Sometimes the water-solubility may be depressed by adding a miscible organic solvent such as ethanol or acetone. Also, some excess of the precipitating agent must remain in solution when the precipitation is complete. This depresses the solubility of the precipitate by the 'common ion' effect.

(b) The precipitate must be uncontaminated by adsorbed impurities.

The conditions are arranged so that the particle size is as large as possible. This reduces the surface area per unit weight of precipitate and hence the opportunity for impurities to be adsorbed on the surface of the particles. 'Granular' precipitates such as barium sulphate are satisfactory from this point of view. Gelatinous precipitates, such as aluminium hydroxide, are extremely unsatisfactory.

As a general rule, unless there are unusual solubility considerations, high temperature favours large particle size. The precipitant, which should be in dilute solution, must always be added slowly with stirring, to avoid a high local concentration. 'Ageing', or allowing to stand for a period before filtration often helps.

(c) The precipitate must be transferred without loss to the weighing vessel.

A filter paper or filter crucible of a porosity fine enough to retain the precipitate particles must be used. The last traces of precipitate adhering to the inside of the beaker must be dislodged with a glass rod fitted with a flexible rubber or plastic cap. 'Ashless filter paper' must be used if the precipitate is to be ignited.

(d) The precipitate must be heated to constant composition and weight.

The conditions must be established for achieving this. Sometimes drying at 100/110°C will be sufficient to reduce the substance to its stoichiometric composition. On the other hand, a temperature of 1100°C is necessary to convert aluminium hydroxide to aluminium oxide. Then again, an intermediate temperature would be appropriate for converting an organo-metallic complex such as titanium 'cupferrate' to the oxide TiO_2.

Sometimes it is possible to precipitate from a homogeneous solution by producing the precipitant through a secondary reaction. For example the oxalate ion may be formed by hydrolysis of some esters. This technique produces superior precipitates.

In addition to precipitation procedures, there are a number of other analytical tests which may be properly regarded as being gravimetric in so far as the final measurement is a weighing. These include:

The determination of the non-volatile residue in a volatile liquid.

The determination of the water insoluble matter in a water-soluble salt.

The determination of silica by weighing before and after treatment with hydrofluoric acid.

7

SPECTROSCOPIC ANALYSIS

7.1 PRINCIPLES

Many modern methods of analysis depend upon measuring the absorption or the emission of radiation energy by the sample. By radiation energy we are including visible light, ultra-violet light and infra-red radiation. These are the most widely used. Other forms of *spectroscopy* include X-ray spectroscopy and nuclear magnetic resonance spectroscopy (NMR) but these are highly specialised techniques and we shall not discuss them here.

Electromagnetic radiation may be arbitrarily divided into several regions according to the wavelength ranges (Table 2).

The techniques have been developed from some very early forms of analysis which depended entirely on the human eye for observation and measurement. For instance, a coloured substance could be measured by

Table 2

Description	Wavelengths	
X-rays	10^{-12}	$\rightarrow 10^{-8}$ m
Ultra-violet (UV)	10^{-8}	$\rightarrow 3 \times 10^{-7}$ m
Visible light	3×10^{-7}	$\rightarrow 8 \times 10^{-7}$ m
Infra-red (IR)	8×10^{-7}	$\rightarrow 3 \times 10^{-4}$ m
Micro-waves	3×10^{-4}	$\rightarrow 1$ m
Radio-waves	1	$\rightarrow 10^{3}$ m and beyond

comparing its depth of colour with a series of standards. A familiar example was a solution containing the purple permanganate ion.

Another early example was the detection of a metal ion by observing the colour imparted to a flame when a few particles of a salt were introduced.

Spectroscopic methods are refinements and extensions of procedures such as these in which instruments have been developed to both produce and measure the radiation concerned. The methods may be subdivided according to the wavelength of radiation, whether the radiation is being emitted or absorbed by the sample and the form in which the sample is presented to the instrument.

7.1.1 Molecular Absorption Spectroscopy

The sample is a liquid or solution. Light is passed through the sample and a measurement is made of the *radiation absorbed* (the absorbance). We speak of UV, visible and IR molecular absorption spectroscopy depending on the wavelength. The absorbance is brought about by an interaction between the incident radiation and the molecules of the substance being determined.

7.1.2 Atomic Absorption Spectroscopy

The sample is submitted as a fine mist to a flame. Light is passed through the flame and a measurement is made of the radiation absorbed. The absorbance is caused by an interaction between the incident radiation and the atoms present in the flame. The sample may also be converted into atoms electrothermally.

7.1.3 Atomic Emission Spectroscopy

Here, we are concerned with measuring the radiation given out by the sample. Energy is fed to the sample, which may be in solid form, by means of a high energy electrical discharge. This raises the temperature and causes the atoms present to give out light, as in the early example of the flame test for metal ions.

The sample may also be in solution. It is converted to a fine mist and sprayed into a flame. In *plasma spectroscopy* the sample is also submitted to the instrument in the form of a fine mist. This is energised by passage through a Plasma Torch which raises its temperature to about $10\,000°C$. The emitted radiation from the atoms present in the plasma are separated and measured. Over forty different elements may be determined in one analytical operation.

Each of these techniques will now be described in more detail.

7.2 MOLECULAR ABSORPTION SPECTROSCOPY

Many compounds are coloured or form coloured solutions. Others form coloured substances by the addition of a suitable reagent. The measurement of the depth of colour and comparison with colours produced by standards provides the basis of a method of analysis. Molecular absorption spectroscopy provides a precise and accurate method for measuring and comparing colours in liquids and solutions.

Colour is perceived when not all the components of white light are transmitted by the substance. Some colours or wavelengths are absorbed by the molecular or ionic species. For example, the permanganate ion absorbs green, yellow and orange light and transmits red, blue and violet light.

Light absorption occurs when the molecule or ion absorbs radiation energy and promotes one or more electrons to higher energy levels. Among inorganic ions, those containing unpaired electrons behave in this way. We may compare the sodium Na^+ ion with the copper Cu^{2+} ion. The former has a completely paired inert gas structure and the latter contains an unpaired electron. Sodium salts do *not* absorb light and are colourless but copper (II) salts *do* absorb light and appear blue/green.

Among organic compounds, those which are 'electron deficient' or 'unsaturated' tend to absorb, producing coloured transmitted light. This is particularly likely when a double bond (4 electrons) is formed between carbon and an element other than carbon, e.g. nitro compounds which contain the group $-\underset{\underset{O}{\|}}{N} \to O$. Such groups are called *chromophores*.

7.2.1 Beer's Law

If a beam of monochromatic radiation (single wavelength) is passed through a coloured solution or liquid contained in a rectangular transparent cell, its behaviour may be described in the following way:

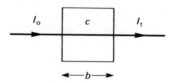

Let I_0 = intensity of incident beam, I_t = intensity of transmitted beam, c = concentration of the absorbing species either as g/dm^3 or mol/dm^3, b = path length in cm.

The percentage of light transmitted by the cell

$$= \frac{I_t}{I_0} \times 100\% = \% \, T$$

The colour intensity is given by the *absorbance* (A)

$$A = \log_{10} \frac{I_0}{I_t} = Ebc$$

It is proportional to the path length b, the concentration c and a constant E which is the *Absorptivity*. If c is expressed as g/dm^3 E is the *specific absorptivity*; if expressed as mol/dm^3 it is the *molar absorptivity* (E_m). In addition, it is often convenient to speak of the $E_{1\,cm}^{1\,\%}$ value. This is the absorbance calculated for a concentration of 1% w/v in a 1 cm path length cell. Deeply coloured substances have large E values.

If b is held constant by using a cell of fixed path length, then $A \propto C$, i.e. absorbance is proportional to concentration (figure 9). This relationship forms the basis of all quantitative methods of molecular absorbance spectroscopy. In practice there are often deviations and some of these are described later.

The instrument used for measuring absorbance is a *spectrophotometer* (see Plate 5). Most instruments cover the whole of the visible spectrum (400–750 nm) extended into the ultra-violet at 200 nm and infra-red at 1000 nm (1 nanometre (nm) = 10^{-9} metres.) A different instrument is used for longer wavelengths in the infra-red region and this is described separately.

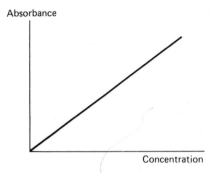

Figure 9   *Calibration curve obeying Beer's law*

A tungsten filament lamp provides a beam of white light. This is dispersed into its separate wavelengths by a prism or a diffraction grating (the monochromator). A very narrow waveband, approaching truly monochromatic light, is selected for its absorbance by the substance being measured and passed through an exit slit. The selection mechanism is geared to a control which indicates the selected wavelength on a calibrated scale.

The selected light is passed through the solution or liquid under test contained in a glass cell. The transmitted beam is passed to a photomultiplier detector which responds by producing a signal proportional to the intensity of the light impinging upon it. The value may be presented either as percentage transmission or as an absorbance. It is usually expressed relative to water or a reagent blank contained in an identical cell (the reference cell).

Most modern instruments are of the 'double-beam' type, which ensure that any instrumental variations apply equally to the sample and the references. Typically, the selected beam from the monochromator is split into two. One 'half' passes through the sample cell and the other 'half' passes through the reference cell. Each 'half-beam' is 'chopped' so that it falls alternately on the photomultiplier detector which develops an alternating current. The waveform signal is converted into two voltages proportional to the light intensities of the sample and reference beams. The voltages are fed to a circuit which converts their ratio to a value which is proportional to the absorbance. The absorbance is displayed on a meter.

In addition to measuring the absorbance at a selected wavelength, it is possible to plot the change in absorbance with the wavelength of the incident beam, using a recording spectrophotometer. The graph so obtained is the *absorption spectrum* (figure 10).

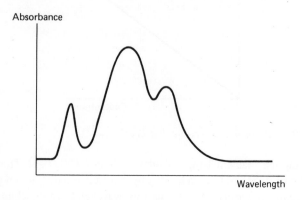

Figure 10   *A typical absorption spectrum*

A particular absorbing molecular or ionic species produces a characteristic absorption spectrum from which it can be recognised. Normally the light chosen for a quantitative measurement will have a wavelength corresponding to a peak maximum in the absorption spectrum. This will produce maximum sensitivity. In some cases, another absorbing species may be present. By comparing the absorption spectra for the two species it is often possible to select a wavelength at which the unwanted species will not interfere.

7.2.2 Ultra-violet Range

Many substances (mainly organic) which do not absorb in the visible range and are therefore colourless do absorb in the ultra-violet range of 200–450 nm. A tungsten filament lamp is not capable of providing the required light; instead, a deuterium discharge lamp is used. For examining such substances, an organic solvent is needed which does not absorb in the required wavelength region. Glass absorbs UV light and cannot be used for the container cell, which is manufactured from clear silica.

7.2.3 Deviations from Beer's Law

For exact compliance several conditions have to be satisfied.

(1) The incident beam must be monochromatic and parallel.
(2) The sample must be a true molecular or ionic solution.
Particulate or colloidal particles cause light scattering.
(3) The solution must be sufficiently dilute to avoid changes in molecular structure within the concentration range being studied.

The incident beam may fail to be monochromatic for two reasons.

(1) It may consist of a wide waveband centred about the required wavelength.
(2) It may contain 'stray' radiation of widely different wavelengths arising from dirty lenses, reflective surfaces, etc.

The effect of the first condition is illustrated in the following way (figure 11).

If a wide waveband is used with a narrow absorption peak, the observed value will be less than the true value. The deviation will be greater at higher concentrations, and the result will be a negative deviation from Beer's Law.

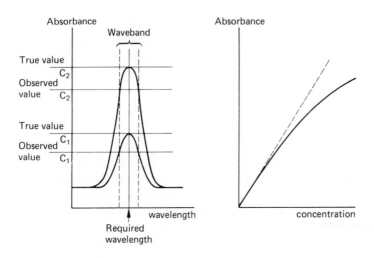

Figure 11   *The effect of wide waveband radiation*

The effect of the second condition may be shown like this. Let I_0 = intensity of required incident wavelength light, I_t = intensity of stray wavelength light, S = intensity of transmitted required wavelength light. Let us assume that the stray wavelength light is transmitted without being absorbed.

Then the measured absorbance value will be:

$$A_{apparent} = \log_{10} \frac{I_0 + S}{I_t + S}$$

As the concentration approaches infinity, the absorbance should also approach infinity as I_t approaches zero. But instead:

$$A_{apparent} = \log_{10} \frac{I_0 + S}{S}$$

Let us suppose that S is 1% of I_0. Then the apparent absorbance will only approach $\log_{10} 101 \simeq 2$. Again there will be a negative deviation from Beer's Law.

These undesirable effects may be minimised as follows.

(1) Use as narrow a waveband as possible.

(2) Keep the glass or silica cells spotlessly clean.

(3) Work wherever possible at a wide rather than a narrow absorption peak.

(4) Have the optics of the instrument regularly serviced.

In practice, it is always preferable to calculate the absorbance values by reference to a calibration curve prepared under exactly similar experimental conditions and covering the appropriate range of concentrations.

7.2.4 Infra-red Range

Beyond the red end of the visible spectrum in the direction of increasing wavelength is the infra-red range of electromagnetic radiation.

Just as coloured substances absorb visible light and some substances absorb ultra-violet radiation, most organic and some inorganic substances absorb infra-red radiation at certain wavelengths. The analytically useful part of the infra-red spectrum extends from 1 μm to 25 μm (μm = 10^{-6} metres).

It is more usual to describe IR radiation by its *Wave Number* than by its wavelength. It is the number of wavelengths per centimetre of radiation

$$\text{Wave Number } v = \frac{1}{\text{Wavelength } \lambda}$$

λ is expressed in cm and v in cm^{-1}.

Although it is quite possible and sometimes desirable to measure IR absorbance at a chosen wavelength as in visible and UV molecular absorption spectroscopy, it is much more usual to plot a continuous IR absorption spectrum. A recording IR spectrophotometer is used. This is a 'double beam' instrument in which the incident beam is split as in the recording visible/UV instrument.

The detector receives transmitted radiation alternately from the sample and reference cells. A difference in intensity causes an AC signal to be produced. This signal is amplifed, rectified and applied to a DC servo motor. The motor opens or closes a variable aperture in the reference beam until the intensities are balanced and the AC signal is zero. The movement is transmitted to the drive of a pen recorder which records percentage transmission or absorbance verses wavelength change.

The optical components of an IR spectrophotometer differ from those in a visible/UV instrument. Firstly the radiation source is a heated ceramic rod often made of sintered oxides of zirconium, thorium or cerium. This produces a continuous spectrum of radiation.

Secondly, glass cannot be used for the monochromator prism or for the sample cells because it is opaque to IR radiation. Alkali halides are used instead: sodium chloride, potassium bromide or occasionally caesium iodide.

Thirdly, the detector is a thermopile rather than a photomultiplier detector.

The sample may be presented as a very thin layer in a short path length cell or as a disc incorporated with potassium bromide. Each molecular species gives a characteristic IR absorption spectrum. Within this spectrum, absorbance peaks may be finger-printed with certain atomic groups as in the illustration of the IR spectrum of 4-methyl-acetophenone (figure 12). Thus this technique may be used as a method of identification and for the detection of impurities containing atomic groups not present in the major component. With a very few reservations, identical IR spectra infer identical substances.

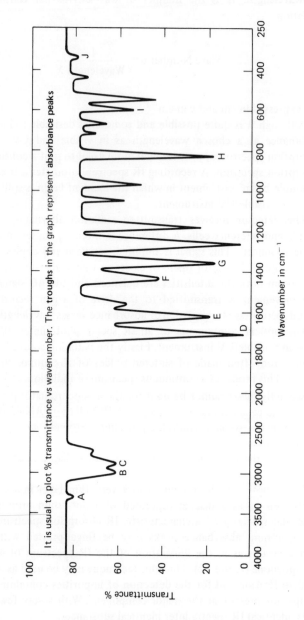

Figure 12 *The infra-red spectrum of 4-methyl-acetophenone*

The absorbance peaks are caused by the following groups

A ketone $C=O$

B aromatic $C-H$

C CH$_3$ linked to $C=O$

D aromatic ketone $C=O$

E aromatic $C=C$

F methyl $-CH_3$

G methyl ketone $-COCH_3$

H 1:4 substitution in benzene ring

I methyl ketone $-COCH_3$

J CH$_3$ linked to $C=O$

Figure 12 (continued)

Unlike visible and UV light, infra-red radiation does not bring about electron transfer from low to high energy levels. Its mode of action is to accentuate the inter-atomic movements within a molecule. For example, a molecule represented as A—B may vibrate along its axis at a particular frequency. Infra-red radiation of the same frequency (and hence characteristic wave number) will be absorbed by the molecule to increase the amplitude of vibration. Other interatomic movements including 'stretching' and 'bending' also have characteristic frequencies and are able to absorb IR radiation of appropriate frequency and wave number. Groups such as $-OH$, $-NH_2$, $>C=O$ each absorb in the same wave number region independently of the remainder of the molecule and may be recognised as peaks in the infra-red absorption spectrum. The technique is often used to elucidate the molecular structure of a new substance.

7.3 ATOMIC ABSORPTION SPECTROSCOPY (AAS)

In molecular absorption spectroscopy the sample is contained in a transparent cell as a liquid or solution, and the chosen radiation is passed through it to the detection system. In atomic absorption spectroscopy, the cell is replaced by a flame and the sample solution is introduced in the form of a fine mist.

The high temperature of the flame causes three things to happen:

(1) The solvent (usually, but not always water) evaporates.

(2) The molecules and ions present in the solute dissociate into atoms to produce an atomic vapour.

(3) A small proportion of the atoms are excited; electrons are promoted to higher energy levels. These often impart a colour to the flame due to the emission of energy when the promoted electrons return to their ground states.

The large majority of the atoms remain in their ground (unexcited) states. If we pass a beam of light through a flame populated by ground state atoms, some of these atoms will absorb certain wavelengths and its intensity will be thereby reduced. This is the principle of the technique. It is used widely for the determination of trace elements in their atomic states.

Suppose, for example, we wish to determine calcium as an impurity in sodium chloride. We prepare an aqueous solution of the salt and introduce it into a flame. The flame becomes bright yellow due to the radiation given out by the excited sodium atoms. We then pass through the flame a beam of light produced by excited calcium atoms. The frequency (energy) of this light is in resonance with the frequency (energy) of the potential electron transitions of the calcium atoms present in the flame. Some of the energy is therefore absorbed by the calcium atoms with the result that the light intensity is diminished during its passage through the flame. The higher the concentration of calcium atoms, the greater will be their absorbance of incident light. A measurement of this value enables us to calculate the calcium content of the sample.

The beauty of the technique is that the chosen light will be absorbed only by unexcited atoms of the species which produced it. For this reason, other atoms do not interfere, for example the huge excess of sodium atoms in the example just described. The preparation of the analytical solution is therefore simple, requiring no elaborate separation techniques.

The instrument is called an *atomic absorption spectrophotometer* (see Plate 6) and its operation may be illustrated schematically by reference to figure 13.

The incident radiation is produced by a *hollow cathode lamp*, in which the cathode is either made of or coated with the metal in question. The surface is bombarded by electrons causing excitation of the chosen atoms and their consequential emission of characteristic radiation. The output from the lamp is *modulated*, i.e. converted to an alternating signal and passed through the flame. The flame is of long path length to give maximum sensitivity to low concentrations.

The analytical solution is converted into a fine mist by passage through the *nebuliser*. The solution meets a mixture of fuel gas and air moving at right angles to its direction of flow, which breaks it into fine droplets. The larger of these coalesce and are drained away. The remaining mist of very small globules is passed into the flame.

The transmitted beam, reduced in intensity, is passed through a *monochromator* (see section 7.2) which enables the selection of a particular wavelength for measurement, and thence to the *detector*. This is tuned to the alternating frequency of the modulated signal from the cathode-ray lamp. In this way, the emitted radiation caused by the excited atoms of the same element in the flame cannot be detected.

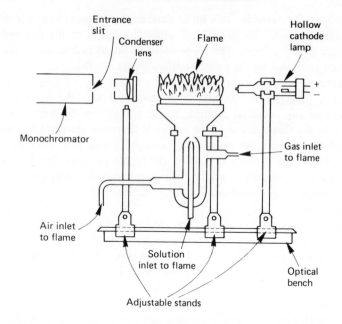

Figure 13 *Atomic absorption spectrophotometer arrangement*

The detector responds by producing a signal which is proportional to the intensity of light falling on it. This is proportional to the concentration of unexcited atoms in the flame which in turn is proportional to their concentration in the analytical sample. The signal is amplified and processed to produce a direct reading of absorbance or percentage transmission.

The terms have the same meaning as in molecular absorption spectroscopy described in section 7.2.

The instrument is first adjusted to give a reading of zero absorbance with a blank solution aspirated into the flame. This should be identical to the analytical solution except for the absence of the element being determined. For example, if it is necessary to dissolve the sample in dilute hydrochloric acid, then acid of the same concentration should be used as the blank. The analytical solution is then substituted and the absorbance reading noted relative to the blank. A second portion of the analytical solution is then taken and a *standard addition* is made of the element being determined. The absorbance is again noted. The increase in absorbance is due to the standard addition. Hence the concentration of element responsible for the first absorbance value may be calculated by simple proportion (Beer's Law, Absorbance $\propto$ concentration, operates as in molecular absorption spectroscopy). As an alternative to the standard addition method, the sample reading may be referred to a calibration curve prepared by subjecting standards to an identical procedure.

There are some cases in which other elements can interfere. For example:

(1) Phosphate ion suppresses the absorbance of calcium atoms. This is probably due to the formation of refractory calcium phosphate which is not dissociated into its constituent atoms in the flame. The addition of salts such as lanthanum chloride can reduce this effect presumably by preferential combination of the lanthanum ions with the phosphate ions, thus freeing the calcium ions for maximum absorbance.

(2) High concentrations of lighter alkali metal atoms cause the suppression of the absorbance of potassium. The effect is thought to be due to the ionisation of some of the potassium atoms in the flame causing a change in its absorbance characteristics. The effect may be countered by adding a caesium salt.

$$K^+ + Cs \cdot \rightarrow K \cdot + Cs^+$$

(3) The absorbance of magnesium is suppressed by the presence of aluminium, silicon, titanium and zirconium. This is probably due to some of the magnesium atoms being incompletely converted to atomic vapour in their passage through the flame. A combination of release agents and a higher temperature flame is often an effective remedy.

The sensitivity of AAS depends on the metal being determined and the wavelength selected for measurement. Each line has its own absorbance characteristic just as in molecular absorption spectroscopy, each absorbing species has a characteristic absorbance. Many elements may be determined at levels of below one part per million in the analytical solution.

7.4 ATOMIC EMISSION SPECTROSCOPY (AES); THE PLASMA TORCH

This was the first form of spectroscopic analysis to be developed.

At normal temperatures the electrons in an atom are in levels of relatively low potential energy. The atom is said to be in its *ground state*. If this atom is supplied with energy, for example from a flame or a high voltage discharge, some electrons absorb energy and are promoted to levels of higher potential energy further from the nucleus. They are unstable in this situation and return to their ground states. In so doing, they emit energy in the form of electromagnetic radiation, much of which falls within the visible and UV spectrum.

For a given type of atom, the possible transitions between ground state and higher potential energy levels are limited and specific for that atom. The wavelength of a particular energy emission depends on the difference in potential energy between the excited and ground states. The greater the difference, the smaller the wavelength and the higher its energy.

The sum total of emitted radiations is the *atomic emission spectrum* of the element. A low energy source, for example a gas–air flame, will not initiate many transitions and a partial emission spectrum will be produced. This may be quite sufficient for some purposes. The higher the source of energy the more complete will be the emission spectrum.

7.4.1 The Sample and the Energy Source

Gases: These are filled into a transparent vessel called a *Geissler tube* and excited by a high voltage discharge.

Solids: These may be powdered and placed in a cavity formed in the end of a graphite rod which acts as an anode. A cathode of pure graphite is clamped at a suitable distance above the anode and a high voltage discharge is passed across the gap. The anode and cathode functions may be reversed.

   Alternatively, if the sample is in rod form, as with a metal sample casting, the sample may itself form the electrode.

Solutions: These may be aspirated into a flame as in atomic absorption spectroscopy.

7.4.2 The Instrument and the Detection System

Part of the emitted radiation passes into the *collimator* via an *entrance slit* from which it emerges as a parallel beam. The beam is *dispersed* by a prism or *diffraction grating* into its individual wavelengths. These may be brought to focus on a photographic plate to form the complete spectrum. This consists of a series of images of the entrance slit (lines), each line representing a single wavelength of radiation. The plate is calibrated so that the individual lines may be identified.

The sample spectrum is compared with the spectrum of a standard substance or those of authentically pure elements.

The *intensity of a line* on the photographic plate is related to the quantity of element producing it. The relative intensity or optical density of the line may be measured with a *microphotometer* or *densitometer*. A beam of white light is passed through the spectrum plate background and again through the line. A photoelectric cell responds to each signal and the signal difference gives the optical density.

In modern instruments, individual lines are focused on separate detectors and their intensities automatically compared with standards to give direct readings. The *direct reading spectrograph* is more fully described shortly.

The advantages of traditional atomic emission spectroscopy are:

(1) The analysis is rapid and often requires very little sample preparation.
(2) Many elements in a complex sample may be studied simultaneously. It is an excellent method of qualitative analysis.

However, it has serious disadvantages as a quantitative method:

(1) The precision and accuracy are not generally good.
(2) The intensity of a given emission is dependent on the other species present. This is the *matrix effect*.

Atomic emission spectroscopy is often used for the specific determination of an element in cases where the technique presents advantages over atomic absorption spectroscopy. Very often, an AAS instrument is modified to accommodate the technique. The sample solution is aspirated into the flame as in AAS. The emitted radiation is mechanically chopped to produce an AC Signal. This is passed to the monochromator. The selected wavelength intensity is measured from the signal produced by the detector which is tuned to the same frequency.

In section 7.3 we considered atomic absorption spectroscopy (AAS) and in the present section, traditional atomic emission spectroscopy (AES). Even if we are equipped with modern versions of both techniques, we shall be aware of certain shortcomings.

AAS requires the separate determination of each element. This means that we need a large number of hollow cathode lamps. The determination of many elements becomes tedious.

Both AAS and Solution AES are limited in their sensitivities. Concentration techniques often have to be used which lengthen the analytical procedures.

7.4.3 Plasma Emission Spectroscopy

This technique enables us to determine many trace elements in one analytical operation (see Plate 7). It is usually much more sensitive, which means that we can determine much smaller quantities without a preliminary concentration step. It is also a more precise technique which means that the spread of a series of replicate values determined on the same sample is less than for the other forms of atomic spectroscopy.

The analytical sample must be in solution form. If the substance is not soluble in water, it is dissolved in a suitable acid such as hydrochloric acid and the solution is diluted to volume in a calibrated flask.

The solution is introduced into the nebuliser which converts it into a fine mist, as in atomic absorption spectroscopy. A *peristaltic pump* may be used to force the solution into the device so that the viscosity of the solution does not impair its flow. The sample mist is then swept into the *plasma torch* (figure 14).

In one version this consists of three concentric quartz tubes. The nebulised sample is carried by a stream of argon up the inner tube. Pure argon (the plasma gas) is passed up the second tube. Nitrogen is passed up the outer tube to form a cooling blanket without which the whole device would melt at the very high temperature produced.

Wound around (but not touching) the open top of the device is the *induction coil*. It is made of hollow copper tubing and is water-cooled. Typically it carries a radio frequency current of about 27 megahertz (MHz) at 5 kilovolts and 1 amp (5 kilowatts). The current is produced by a specially designed power unit for high stability. The energised coil acts as the primary winding of a transformer.

The argon 'plasma gas' behaves as the secondary winding. It is ionised and hence made conducting by initially passing a high voltage spark.

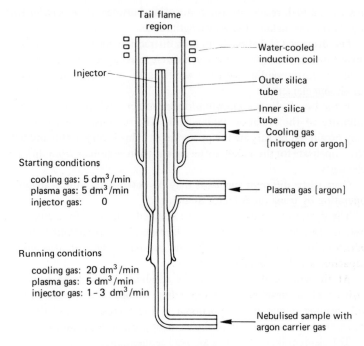

Figure 14   *The plasma torch (one version)*

The high input of energy produces a temperature of up to about
10 000°C in the gas volume contained by the upper part of the device
encircled by the induction coil. All the elements present in this 'plasma'
are excited. Electrons are promoted to higher energy levels and are
often eliminated altogether from the mother atom to give an ion. A
very complex and complete emission spectrum of the test sample is produced consisting of both atomic and ionic lines. We often make use of
the ionic lines; they frequently permit greater sensitivity.

The emission is rich in the ultra-violet range. For this reason, the
plasma torch is fitted behind a safety door with a radiation-absorbing
window.

Part of the plasma emission is collected from the 'tail flame' region
just above the open end of the plasma torch by a lens which converts it
into a parallel beam. The beam enters the *direct reading spectrograph*
via an entrance slit and is divided. Each beam is separately dispersed by
a diffraction grating. One grating handles UV wavelengths and the other
carries on through the visible range. By dividing the total spectrum in

this way, a high resolution (discrimination between close wavelengths) is achieved throughout the wavelength range.

The dispersed radiation from each diffraction grating is focused on a curved metal strip called a *Rowland Circle*. Slits are cut in this device at positions corresponding with the wavelengths of selected emission lines for chosen elements.

Behind each slit is a *photomultiplier detector* which responds to the intensity of the radiation falling on it. The response of each detector may be altered by varying the power input. This is very useful and helps to compensate for the widely differing emission responses from different elements.

Over forty different elements may be determined in one analytical operation by using the Selector Switches.

The detectors are exposed to the selected radiations for a given time period, usually 15–30 seconds. Each responds by producing a current which persists for the duration of the exposure. The current is fed to a capacitor which accumulates a charge.

At the end of the exposure, the condenser is discharged across a resistor which develops a corresponding voltage. The integrated millivolt reading is directly proportional to the intensity of radiation and hence to the concentration of the element present in the sample.

The element line intensities are read sequentially.

A multielement standard solution containing known concentrations of the chosen elements in a similar environment is run under identical experimental conditions to that of the standard. The analytical results are calculated from the readout for sample and standard.

It is possible to feed the readout signals directly into a microcomputer, suitably programmed, and to arrange for the analytical values to be reported and displayed directly in terms of concentration, for example as parts per million.

The technique has the following advantages:

It is extremely rapid, and many elements may be determined at the same time.
The experimental procedure, once established, is easily followed.
It does not require a great degree of experimental skill.
It is very sensitive; most elements may be determined at levels well below 1 part per million in the analytical solution.
The precision is better than 1% relative, in most cases.
The response is linear with respect to concentration over many orders

of magnitude. This means that multielement analysis is feasible and a small number of multiple standards will usually be sufficient.

The chief disadvantage is its cost which is far greater than, for example, an atomic absorption spectrophotometer.

It does have a further disadvantage which is common to all forms of emission spectroscopy and that is the matrix effect already mentioned in connection with traditional AES. The intensity and hence readout for a given concentration of an element is affected by other elements present at higher concentrations. This means that, for example, five parts per million of calcium will not necessarily give the same reading for potassium and sodium chlorides. Different multielement standards must be used in which the major component environment matches that of the sample. However, this effect is much less marked than in traditional AES.

The technique is particularly straightforward for determining trace impurities in organic solvents and mineral acids. The sample is measured and transferred to a platinum basin where it is evaporated to dryness. The matrix effect is thus eliminated. The trace impurities contained in the residue are dissolved in a little hydrochloric acid. The solution is diluted to a known volume and introduced into the plasma torch.

8

CHROMATOGRAPHY

8.1 PRINCIPLES

Chromatography is a method of separation. It may be subdivided according to the physical state of the sample, i.e. liquid or gas, the technique used to bring about the separation and the method used to detect and measure the separated components.

Most methods of chromatography have the following features in common.

The analytical sample is introduced into a moving fluid (*the mobile phase*) and made to pass through a separating medium. This may consist of silica, alumina, paper or glass beads, etc. (the inert support) which has been impregnated with a sparingly soluble liquid, *the stationary phase*. The support may take the form of a packed column, a paper strip, a thin layer on a glass plate, etc.

As the sample is carried by the mobile phase through the separating medium, the individual components are delayed to different extents by contact with the stationary phase. By considering the chemical and physical properties of the components we wish to separate, the mobile and stationary phases may be chosen to accentuate this effect.

The result is that the components are separated by their passage through the separating medium, when they may be detected and measured. The relative degrees of separation of the components of a mixture may be expressed by comparing their *Rate Factors* or *Rf values*.

$$Rf = \text{Ratio of } \frac{\text{Rate of movement of the solute}}{\text{Rate of movement of the mobile phase (solvent)}}$$

8.2 THIN LAYER CHROMATOGRAPHY AND ELECTROPHORESIS

The separating medium consists of an even layer of absorbent such as alumina or silica gel coated on to a glass plate. The absorbent combines the properties of support and stationary phase.

A spot of sample solution is introduced on to the plate near one edge. The plate is suspended in a vertical position with this edge dipping into a solvent mixture, the mobile phase. The operation is carried out in a closed tank so that the atmosphere surrounding the plate is saturated with the vapour of the solvent.

The solvent rises up the plate by capillary effect. The components of the sample solution have different Rf values. They travel different distances up the plate and are separated from each other.

They are detected by spraying the plate with a solution with which they form coloured compounds. In some cases this is unnecessary because the components are themselves coloured, as in the case of dyes. An individual spot may be identified by running a standard on the same plate and comparing the Rf values. Of course, this only works if we have a strong suspicion of its identity in the first place.

The processes involved are not simple and for this reason the experimental conditions usually have to be determined empirically.

The movement of a solute is related to:

The absorptive capacity of the stationary phase for the solute.
The absorptive capacity of the stationary phase for the solvent.
The solubility of the solute in the solvent.

With alumina plates, the absorptive capacity is related to mobile hydroxyl ions or chloride ions (present as an impurity). With silica plates the behaviour is thought to be more closely related to mobile water molecules.

Thin layer chromatography is used extensively for the separation and detection of components in a mixture which could not be identified by purely chemical methods.

8.2.1 Electrophoresis

Some substances, notably amino acids, are *ampholytic*, i.e. they can behave as acids or bases depending on the pH of the aqueous medium in which they are dissolved. For example, in alkaline conditions:

$$R.CH_2 < \begin{array}{c} NH_2 \\ COOH \end{array} \xrightarrow{\ OH^- \ } R.CH_2 < \begin{array}{c} NH_2 \\ COO^- \end{array} + H_2O$$

In acid conditions:

$$R.CH_2 < \begin{array}{c} NH_2 \\ COOH \end{array} \xrightarrow{\ H_3O^+ \ } R.CH_2 < \begin{array}{c} NH_3^+ \\ COOH \end{array} + H_2O$$

Thus a cation (positively charged) or an anion (negatively charged) may be formed. This behaviour may be used to separate, for example, a mixture of amino acids, or proteins.

A strip of absorbent paper is dipped at either end into a compartment containing an electrolyte buffered to a selected pH. A spot or line of sample solution is placed at a pre-determined position towards the end of the paper. A potential difference of about four volts per centimetre length of paper is then applied between the opposite compartments (figure 15).

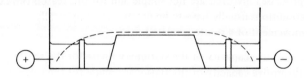

Figure 15   *Electrophoresis*

The charged ions move towards either the anode or the cathode at different rates and are thereby separated. As in thin layer chromatography they may be identified by spraying with an appropriate colour-forming reagent. Standards may be run at the same time on parallel paper strips.

8.3 HIGH PERFORMANCE LIQUID CHROMATOGRAPHY (HPLC)

In the earlier forms of liquid chromatography the separating medium, with or without an added stationary phase, was held in a relatively large cylindrical column. The sample was introduced into the top of the column. The mobile phase was introduced and carried the constituents of the sample downwards by gravity. These were separated, eluted one by one, identified and measured. It was a very slow process.

In recent times, the technique has been made much more efficient.

The columns are of much smaller internal diameter, about 2.5 mm. The column packing is of small particle size, about 3.5 μm.

New and improved substances have been developed as stationary phases.

A high pressure is used to drive the mobile phase containing sample constituents through the column.

The flow rate of the mobile phase may be precisely controlled.

Very small samples may be used.

Special continuous detectors are available capable of handling small flow rates and detecting very small amounts.

Automated instrumentation has been developed capable of rapid analysis and high resolution of the eluted components.

The pump propels the mobile phase.

A device is required to introduce the sample.

The column containing the stationary phase performs the separation.

The detector determines the extent of the separation and enables a quantitative analysis to be achieved.

8.3.1 The Mobile Phase

This plays a vital role in the separation process and it is essential to make an appropriate choice. It may be a single substance or a mixture of substances. The composition may be maintained constant throughout the separation (*isocratic operation*). It may also be systematically changed (*gradient elution*). This is useful if the sample contains constituents of widely differing polarities (distribution of electrical charge within the molecule). A wide range of liquids is available, each one low in impurities likely to interfere with the analysis.

8.3.2 The Stationary Phase

This may be a porous solid of selected chemical composition and particle size. Where an additional stationary phase is required, this is generally bonded to the support particles. Typically, a chemical reaction takes place between the surface hydroxyl groups of hydrated silica atoms and an organic molecule.

8.3.3 Flow Control

A pump provides either constant inlet pressures or constant flow rates. Separation is more efficient at a low flow rate.

8.3.4 Sample Introduction

The simplest method is to inject through a septum with a micro syringe. A difficulty is caused by the high pressure of the mobile phase. Alternatively, sampling devices incorporating valves may be used. Solid samples may be dissolved in a suitable solvent. A very small sample varying from a few nanograms (10^{-9} g) to about 2 mg are employed.

8.3.5 Columns

A typical column is 25–50 cm long of internal diameter from 2 to 5 mm. The packing consists of very small particles of 5–15 μm diameter. Pre-packed columns are generally used.

8.3.6 Column Temperature

The column may be used at ambient or raised temperature. A raised temperature:

Reduces the viscosity of the mobile phase thus lowering the 'back pressure'.

Increases the solubility of the sample constituents in the stationary phase.

Increases the solubility of the sample constituents in the mobile phase.

These goals may generally be achieved at temperatures of 50–70°C.

8.3.7 Detectors

Optical detectors are the most frequently used. A beam of light is passed through the emergent liquid stream as it flows through a transparent cell of very low volume (about 10 microlitres). The incident beam suffers attenuation of one or more kinds in its passage through the cell. The degree of interference will change as the moving phase carries the separated components through the cell one by one. These changes cause variations in the output voltage from the detector which responds to the emergent beam. The output is amplified and made to drive the pen of a strip chart recorder. Alternatively, the signal may be fed into a computer.

We may monitor the following changes:

UV absorbance.
Fluorescence emission.
Refractive index.

In a typical example, the chromatogram traced on the recorder will contain several UV absorbance peaks, each one corresponding to an individual component in the original sample. These may be identified by comparing their retention times (the time which elapses between the injection of the sample and the passage of the component through the detector cell) with those of standards processed in identical experimental conditions. Quantitative analysis may be accomplished by comparing the peak area of the sample with that of a standard (see figure 16: the presentation is similar).

8.4 GAS-LIQUID CHROMATOGRAPHY (GLC)

As its name suggests, in this form of chromatography the moving phase is a gas and the stationary phase is a liquid. The technique enables us to separate, identify and measure the components of a mixture where these can be vaporised at temperatures up to about 300°C. Only a very small sample is required. The instrument is called a *gas chromatograph* (see Plate 8).

A few microlitres (μl) of sample is injected from a graduated micro syringe through a self-sealing cap into a stream of carrier gas. This is usually nitrogen. The injection temperature is sufficient to quickly vaporise all the sample components.

The carrier gas carries the vaporised sample into the *column*. This is a tube of uniform diameter usually of coiled construction so as to accommodate as great a length as possible in a small space. It is packed with an inert siliceous substance (*the support*) which has been impregnated with the stationary phase. This is a high boiling liquid, with very low volatility at the temperature of the experiment. The column is heated in an oven at a sufficiently high temperature to maintain each of the components in their vapour states.

As the vapour mixture passes through the column, the individual components travel at different speeds. This is due to a combination of two effects:

A difference in atomic mass. A light low boiling component travels faster than a heavy high boiling component.

A difference in the solubilities of the components in the stationary phase. Components with a high solubility will travel more slowly because their partition between the stationary and mobile phase will be weighted towards the stationary phase.

If the experimental conditions are chosen correctly, the components will be separated by the column and will emerge one by one, separated from each other by pure carrier gas. They are swept into the detector.

The detector responds to the passage of each component by producing an electrical signal. This is amplified and processed to move the pen of a chart recorder, to produce a *chromatogram* (figure 16). This presents a picture of the composition of the sample.

The figure indicates the presence of four components. X represents the moment of *sample injection*. The distances XA, XB, XC and XD represent the *retention times* of the components.

For a given set of experimental conditions (i.e. stationary phase, column temperature, gas flow rate, etc.) the retention time is a characteristic for a specific substance. It is thus possible to identify the components by comparing the retention times with those for known substances. Very often, an identity is verified by adding a little of the suspect substance to a portion of the sample. If the subsequent chroma-

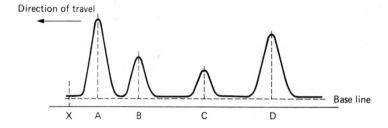

Figure 16 *A gas chromatogram*

togram displays a larger component peak and no additional peaks, the identity is confirmed.

The area of each peak is related to the quantity of component. An approximate measure of the percentage of a component may be obtained by measuring and calculating the relative peak area. This is the area of the component peak expressed as a percentage of the sum of all the peak areas. This is as far as we can go towards quantitative measurement unless we know the identities of the substances present.

Modern instruments are fitted with electronic integrators. These automatically present the peak areas as numerically related values and as relative peak area percentages. Facilities are provided for selectively omitting minor components in the percentage calculations. This enables ratios to be calculated for multicomponent mixtures whilst ignoring incidental impurities.

For accurate quantitative work, the component peak must be compared with a calibration curve showing the change in peak function with concentration. Several peak functions have been recommended. For example:

The peak area relative to that of an internal standard.
The peak height relative to that of an internal standard.
The peak height multiplied by the peak width (at half the peak height) relative to that of an internal standard.

The internal standard is a substance not present in the sample. It is added at the same concentration to each of a series of calibration standards. The peak area ratios are plotted versus concentration to produce the calibration curve. The same concentration of internal standard is then added to the sample and the peak area ratio is referred to the calibration curve.

8.4.1 Detectors

The earliest type was the *katharometer* or *thermal conductivity detector* (figure 17). It is still useful for some purposes.

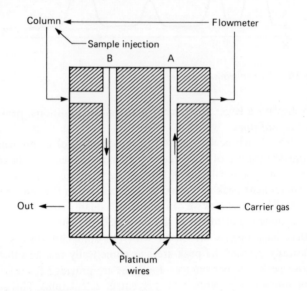

Figure 17 *The katharometer detector*

Two holes of identical dimension are drilled in a brass block. A very fine platinum wire is passed through each hole. Each wire forms one of the sides of a Wheatstone Bridge for the measurement of electrical resistance. The platinum wires are heated, and carrier gas is passed through in the manner shown. The heat from each wire is conducted through the gas to the metal block where it is quickly absorbed. The Wheatstone Bridge is balanced so that no current passes across it.

The sample is injected, separated in the column and the components are conveyed by the carrier gas through the second hole in the metal block. The thermal conductivity of the gas surrounding this platinum wire is changed, causing the temperature of the wire to change.

The electrical resistance of a metal wire is related to its temperature. Consequently, the Wheatstone Bridge is thrown out of balance and a current flows across it. The current is amplified and converted to an electrical potential which moves the pen of a chart recorder.

Probably the most widely used device is the *flame ionisation detector* (FID) (figure 18). The carrier gas carrying the separated components

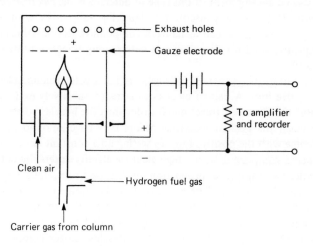

Figure 18   *The flame ionisation detector*

is mixed with hydrogen and air and passed to a burner. This acts as an electrode of earth potential. A positive potential is applied to a metal gauze located above the flame. As each component enters the flame, some of the molecules are ionised by the absorption of energy. This causes a current to flow between the electrodes. The current is again converted to an electrical potential which activates the chart recorder.

Another type of detector is the *ionisation cross-section detector*. A *beta-ray emitter* such as strontium 90 or yttrium 90 is incorporated in the detector cell. The gas entering from the column is thereby ionised. A pair of electrodes provides a potential difference across the cell and an 'ion current' is caused to flow. This current is proportional to the molar concentration of each component as it passes through the detector. It is also proportional to a value called the *specific cross-section*. This value is characteristic for a given species. It depends upon the electronic configurations of the constituent atoms within the molecule and also the molecular mass. The most suitable carrier gases for this type of detector are hydrogen and helium. They have low specific cross-sections and hence produce low background currents giving maximum sensitivity. The detector is only slightly sensitive to temperature and gas flow rate fluctuations. It may be used for the detection of all organic and in-organic substances of sufficient volatility.

A special development of this type of detector is the *electron capture detector*. It is specially useful for detecting substances with a high affinity for electrons such as halogen compounds. In this connection it is frequently used for detecting and determining traces of pesticides in foodstuffs.

The carrier gas, generally nitrogen, is ionised to produce electrons and positive ions. A pair of electrodes attract respectively each type of ion. But the electrons travel much faster than the heavier positive ions. They reach the positive electrode before they have an opportunity to recombine with the positive ions, so setting up a constant ion current.

When a component with a high electron affinity enters the detector, its molecules capture a proportion of the electrons, thus becoming negatively charged ions. These neutralise and combine with the positive carrier gas ions. The ion current is reduced in relation to the concentra-tion. The governing law closely resembles Beer's Law for the absorption of light. This detector is highly selective. Besides halogen compounds, it is sensitive towards nitriles and nitrates and conjugated carbonyl com-pounds. It is almost totally insensitive towards hydrocarbons, alcohols, etc.

In many cases, the column is maintained at a fixed temperature during the analysis. Sometimes, however, one or more of the compon-ents is not sufficiently volatile to be eluted at the selected temperature. *Temperature programming* enables us to raise the temperature in a controlled fashion throughout the analysis. Low boiling volatile com-ponents emerge at the lower temperature. High boiling, less volatile components are hastened on their way by the higher temperatures.

This technique not only ensures that less volatile components are not missed in the analysis. It also produces a chromatogram consisting of sharp, well-defined peaks instead of one in which the peaks become broader and less well-defined as their volatility decreases.

Gas–liquid chromatography has revolutionised the organic chemistry analytical laboratory. In addition to being a complete analytical tech-nique in its own right, it is also used to separate the components of a mixture before determining them by other methods, for example by advanced forms of spectroscopy.

9

POLAROGRAPHY

This technique is based on the behaviour of ions in a specially designed electrolytic cell. For many years it was very popular for the determination of trace metals and other electro-reducible ions. Although atomic absorption and plasma emission spectroscopy have largely replaced it for trace element analysis, it still occupies a useful place in the repertoire of instrumental methods. In fact it is currently enjoying a new lease of life. Unlike AAS and AES, the method permits us to distinguish between different oxidation states, for example chromium (III) and chromium (VI).

If two inert electrodes, say of platinum, are immersed in a stirred aqueous solution of a metallic salt, no current will flow through the cell until the applied potential difference reaches a certain value, depending on the ion in question. At this point, electrolysis commences, the metal ions give up their positive charges and the corresponding metal is deposited on the cathode. At higher potentials, the cell current increases in accordance with Ohm's Law:

$$\text{current} = \frac{\text{voltage}}{\text{resistance}}$$

In polarography, the electrolytic cell (figure 19) is of a special design and, most important, the solution is not stirred.

Mercury is supplied from a reservoir to a glass capillary tube which delivers it at a steady rate. The succession of growing mercury drops forms the *dropping mercury electrode* (DME). This is most commonly the cathode (−ve).

71

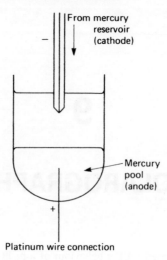

Figure 19 *Polarography cell*

The anode (+ve) is a mercury pool connected to the external circuit by a platinum wire fused into the base of the glass cell. Alternatively it may be a standard reference electrode such as the saturated calomel electrode.

The test solution, containing the ions to be determined, is placed in the cell and a steadily increasing negative potential is applied to the dropping mercury electrode. Facilities are provided for measuring the applied potential and the resultant cell current. The arrangements may be understood by considering a schematic diagram, featuring the mercury pool anode (figure 20).

In modern instruments, in addition to the reference electrode, an auxiliary electrode is incorporated in the assembly. This is made of inert platinum or graphite. The cell current is actually measured between this and the dropping mercury electrode. As in the simple electrolysis cell, no current flows until the applied potential reaches the critical value for the ion in question. At this point the behaviour is different.

Let us consider what is happening at the dropping mercury electrode (figure 21). The positive ions in the immediate vicinity of the drop surface are reduced to the corresponding uncharged atoms which amalgamate with the mercury. Since the solution is not stirred, the effect is to produce a surface layer of solution in which the positive ion concentration is virtually zero.

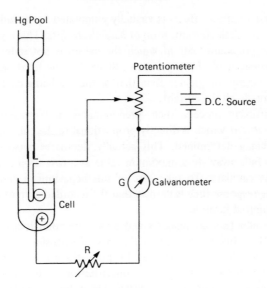

Figure 20 *Polarographic circuit*

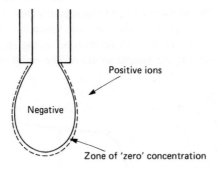

Figure 21 *The dropping mercury electrode*

There are two ways in which this depleted zone may be replenished. By *electrostatic attraction* of the negatively charged mercury drop for the positive ions in the bulk of the solution which gives rise to the *migration current*. By *diffusion of ions* from the bulk of the solution. The rate of diffusion is directly proportional to the concentration of the positive ions in the bulk of the solution.

The first of these effects is virtually eliminated by including in the test solution a high concentration of *base electrolyte*. This is generally a sodium or potassium salt in which the cation is not reduced at the applied potential of the experiment. Since its concentration is several orders of magnitude greater than that of the ion being determined, it carries virtually all the current.

The diffusion process, then, predominates, and gives rise to the *diffusion current* which is directly proportional to the concentration of the ions being determined. This actually fluctuates between zero as each drop falls away, to a maximum value as a new drop grows at the end of the capillary. To take care of this behaviour, a galvanometer with a long response time is used so that the recorded current fluctuates in only a limited fashion.

At V_1 reduction commences and at V_2 the point is reached when the current is directly related to the rate of diffusion and hence the concentration (figure 22). Beyond this point there is little change. The step-height is proportional to the concentration of the reducible ion. V_x represents the *half-wave potential*. The value is characteristic for each reducible species in a given base electrolyte environment.

The relationship is plotted on a moving chart recorder and the curve so produced is the polarogram. The polarograms from the samples are compared with polarograms from standards which have been processed identically. Alternatively, a standard addition of the species is made to a measured volume of the sample solution. The result is calculated by

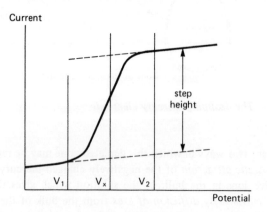

Figure 22 *Current–potential relationship in polarography*

relating the increase in step-height brought about by the addition to the step-height of the sample solution alone.

More than one metal may be determined from the same polarogram provided that the half-wave potentials of the two differ by not less than 0.15 volts (figure 23). When the difference is less, it is sometimes possible to alter the value for one of the species by adding a complexing agent. This converts the simple ion into a complex ion.

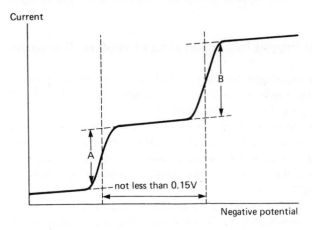

Figure 23 *Polarographic waves for two metals*

Cathode-ray Polarography

In the 'simple' polarography we have been considering so far, the increasing potential difference is applied during the lives of many mercury drops. In a cathode-ray instrument, the potential difference increase is applied towards the end of the life of each drop. In earlier instruments, if the total life span of each drop were seven seconds, the 'potential sweep' was applied during the last two seconds. In modern instruments, this has been reduced to about 20 milliseconds.

The potential sweep and drop rate are synchronised. The potential is applied to the 'X' plates of a cathode-ray tube with a long 'afterglow time'. The cell current, which increases during each sweep, is passed through a load resistor and the potential drop across this is amplified and fed to the 'Y' plates ('X' and 'Y' are synonymous with the 'X' and

'Y' axes of a graph). The result is that once every sweep, a polarogram is traced on the cathode-ray screen. When viewed in subdued light, the trace remains visible between successive potential sweeps. A peak rather than a step is obtained and the peak height is proportional to the concentration of the reducible ion.

Steps 0.1 volts apart may be measured. As in 'simple' polarography, standards or standard additions are run under identical experimental conditions and the peak heights are compared with those of the samples.

Anodic Stripping Polarography Using a Cathode-ray Polarograph

In the techniques we have looked at so far, the dropping mercury electrode has been used. Instead, a single mercury drop may be held stationary at the end of the capillary.

A negative potential is applied to the drop and the test solution is electrolysed for a measured time. This concentrates the reducible ion in the mercury drop. The current is then reversed, the stationary electrode becomes the anode and the metal is 'stripped' into solution again $(M \rightarrow M^{n+} + e^-)$.

The cell current is proportional to the concentration of reducible ion in the mercury drop. Since this is many times greater than its concentration in the original test solution, the sensitivity is greatly increased. For example, lead may be determined at a level of 0.2 nanograms per cm^3.

Recent instrumental developments have centred on the manner of application of potential to the dropping mercury electrode to enhance sensitivity. Electrodes other than the DME are also used.

In addition to the determination of reducible metal ions, polarography may be used for determining many non-metal ions and organic groups. The requirement is simply that they may be quantitatively reduced within the normal range of applied potentials. The process is dependent only on the migration rate due to the concentration difference between the bulk of solution and the dropping mercury electrode surface layer. Hence neutral species may often be determined.

10

AUTOANALYSIS

This is a term which covers in a general sense any analytical process in which part or all of the steps are carried out automatically instead of by the analyst at the laboratory bench. Thus it could be applied to the on-stream automatic sampling and measurement of a continuous chemical manufacturing process.

However, we shall use the term in a specific sense to characterise a particular type of automatic analysis which is especially widespread in hospital laboratories. It was developed to deal with a situation in which large numbers of the same kind of sample need to be analysed for the same component at similar levels. A good example is the determination of sodium and potassium in body fluids. The equipment is an *auto-analyser*.

The samples are introduced sequentially into a continuous flowing stream of carrier solution in such a way that they are separated by sections of 'pure' solution. Air bubbles are drawn into the stream to produce *segmentation*. This maintains the identities of individual samples whilst permitting a high rate of sample throughput.

Successive modules, arranged to suit the order of analytical operations, perform different tasks, for example, heating, dialysis, and final measurement, which is usually spectrophotometric.

Reagent additions are made by merging sample and reagent streams.

The liquid streams are kept moving by using a multichannel peristaltic pump. This is a simple but ingenious device in which the solutions are propelled along polythene tubes by a rotating series of rollers which

77

compress the tubes in the appropriate direction. It is considered to be the vital part of an autoanalyser. It serves three purposes:

The introduction of sample and reagents into the system.
The transportation of solutions through the system at constant speed.
The provision of air bubbles for liquid segmentation.

The arrangement may be understood with reference to figure 24. The sample presentation unit is usually a plate with holes around the circumference to hold sample cups. A sample probe of stainless steel withdraws the sample and presents it to the analytical system. Commonly, the unit has a capacity for a hundred samples.

The samples and reagent solutions are introduced in appropriate flow ratios by selecting tubing of the correct internal diameter.

The processing rate may vary from 20 to 80 samples per hour.

Several samples are generally being processed at a given time.

Standards are inserted in the sample sequence and these provide regular reference points.

Following the completion of the final analytical process and reagent addition, the moving stream conveys the samples and standards through a UV/visible spectrophotometer (see section 7.2) fitted with a *flow-through sample cell*. The absorbance values are measured continuously during a complete sequence and are automatically displayed by the *read-out system*. This may be a moving chart recorder. In this case, the record takes the form of a graph in which the peaks represent samples and standards in pre-arranged sequence.

The analytical values are calculated from the ratios of peak heights for samples and standards.

Recent developments have centred on the design of easily inter-changed modules for flexibility of analytical operations and automatic recording of measurements and results.

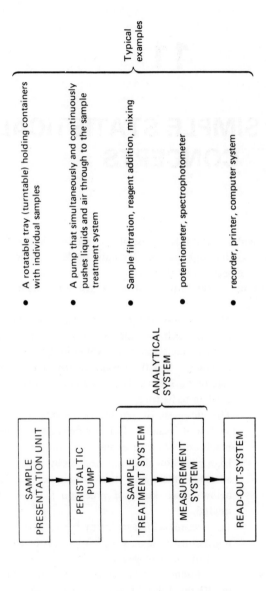

Figure 24 *Autoanalyser arrangement*

Typical examples

- A rotatable tray (turntable) holding containers with individual samples
- A pump that simultaneously and continuously pushes liquids and air through to the sample treatment system
- Sample filtration, reagent addition, mixing
- potentiometer, spectrophotometer
- recorder, printer, computer system

SAMPLE PRESENTATION UNIT

PERISTALTIC PUMP

SAMPLE TREATMENT SYSTEM

MEASUREMENT SYSTEM

READ-OUT-SYSTEM

ANALYTICAL SYSTEM

11

SOME SIMPLE STATISTICAL CONCEPTS

We mentioned in chapter 1 that it is impossible to establish the absolute truth about the composition of a material. We must settle for estimating the composition to the best of our ability.

Let us look at some of the terms used in discussing the validity of analytical results.

Accuracy: The closeness of agreement between the experimental result and the 'true value'. Close agreement is 'high accuracy'; wide divergence is 'low accuracy'. Strictly speaking, the 'true value' cannot be known. It is a standard value which is taken to be true.

Precision: The closeness of agreement between replicate analytical results. Close agreement is 'high precision'; wide divergence is 'low precision'. High precision does not necessarily imply high accuracy; there may be a systematic error (see below). Two kinds of precision may be considered.

Repeatability: The closeness of agreement between replicate results on the same sample by the same analyst using the same reagents.

Reproducibility: The closeness of agreement between replicate results on different samples of the same batch of substance by different analysts using separately prepared reagents.

For a given analytical method, the repeatability is always better than the reproducibility, sometimes by a factor of three or four.

Mistake: A divergence from the 'true value' due to an unintentional departure from the intended procedure.

Error: A divergence from the 'true value' from inherent causes in the intended procedure. These are always present in some form and are not due to the incompetence of the analyst. The word does not have the same meaning as in the general vocabulary. There are two kinds of error.

Systematic error: A divergence from the 'true value' that tends to lean one way. It is not eliminated by averaging replicate results. Sometimes it is called bias.

Random error: A divergence from the 'true value' which, although individually unpredictable, is eliminated by averaging a sufficiently large number of replicate results.

Mean: The average of a number of similarly determined results. It is given the symbol $\overline{x}$ relating to a series of individual results x.

Standard deviation: This is a measure of precision and enables us to express the extent to which replicate results are likely to fluctuate about the mean result.

It is given the symbol s.

It is calculated by the formula:

$$s = \sqrt{\frac{\Sigma (x - \overline{x})^2}{n - 1}}$$

$\Sigma (x - \overline{x})^2$ = the sum of the squares of the individual errors.

n = the number of replicate results.

Let us see how this works out with a practical example. A solution of potassium chloride was analysed by titrating ten successive measured volumes with standard silver nitrate solution. The results are given as grams potassium chloride per 1000 cm^3 solution in table 3.

Table 3

Individual results x	error $x - \bar{x}$	squares $(x - \bar{x})^2$
10.51	−0.10	0.0100
10.57	−0.04	0.0016
10.60	−0.01	0.0001
10.63	+0.02	0.0004
10.59	−0.02	0.0004
10.62	+0.01	0.0001
10.61	0.00	0.0000
10.68	+0.07	0.0049
10.63	+0.02	0.0004
10.61	0.00	0.0000
Mean $\bar{x}$ = 10.61		0.0179
		= $\Sigma (x - \bar{x})^2$

Now $n = 10$ $\therefore n - 1 = 9$

$$\therefore \text{Standard Deviation} = s = \sqrt{\frac{0.0179}{9}} = 0.045$$

The significance of the Standard Deviation may be appreciated by considering the Normal Error Curve (figure 25).

If a very large series of replicate analyses is completed, the results will always fall within the area enclosed by the curve provided that the experimental conditions have not varied within the series.

The total area enclosed by the curve represents 100% of the results.

The percentage of the total enclosed area between any two limits on the horizontal axis indicates the percentage of results which may be expected to occur between these limits.

The area enclosed by one standard deviation [////] represents 66% of the whole area. This means that 66% of a series of replicate results may be expected to lie between the limits $\bar{x} \pm s$.

The area enclosed by two standard deviations [//// plus \\\\\\\\\] represents 95% of the whole area. This means that 95% of a series of replicate results may be expected to lie between the limits $\bar{x} \pm 2s$.

Finally, the area enclosed by three standard deviations, [//// plus \\\\\\\\\ plus ░░░] represents over 99% of the total. Over 99% of a series of replicate results will therefore lie within the limits $\bar{x} \pm 3s$.

Notice that the area enclosed by one standard deviation corresponds with the inflections in the curve.

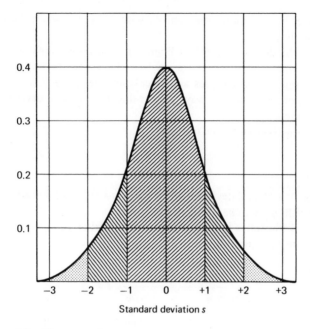

Figure 25 *The normal error curve*

Confidence Limits

A knowledge of the standard deviation of a given analytical method enables us to predict the spread of a series of replicates. It also enables us to quote confidence limits for a single result.

Let us look again at the example. Suppose we carry out just one analysis and the result is, say 10.59. We know that the standard deviation is 0.045.

We can say that the 'mean result' will lie between:

$$10.59 \pm s = 10.59 \pm 0.045 \text{ with } 66\% \text{ confidence}$$

and the 'mean result' will lie between:

$$10.59 \pm 2s = 10.59 \pm 0.09 \text{ with } 95\% \text{ confidence}$$

and the 'mean result' will lie between:

$$10.59 \pm 3s = 10.59 \pm 0.135 \text{ with } > 99\% \text{ confidence}$$

Notice that the more confident we wish to be, the wider the limits we must quote.

It may be that the confidence limits for a single value are not narrow enough for the intention of the analysis. They may be improved by carrying out several analyses and calculating the mean. The standard deviation of the mean is always smaller than the standard deviation for a single result.

The value is given by the formula:

$$s_{\overline{x}} = \frac{s}{\sqrt{n}}$$

where $s_{\overline{x}}$ = standard deviation of the mean, s = standard deviation for a single result, n = number of replicate results.

Table 4 illustrates how $s_{\overline{x}}$ is improved in relation to s as the number of replicates is increased.

Table 4

No. of results n	$\sqrt{n}$	$s_{\overline{x}}$
1	1.000	1.00
2	1.414	0.71
3	1.732	0.58
4	2.000	0.50
5	2.236	0.45
6	2.450	0.41
7	2.646	0.38
8	2.829	0.35
9	3.000	0.33
10	3.162	0.32

By carrying out just four replicate analyses and calculating the mean, the standard deviation may be halved. In our example, if 10.59 had been the mean of four values, the 95% confidence limits would have been not ± 0.09 but ± 0.045.

The greatest advantage is gained by a few replicates. Thus a two-fold improvement may be obtained by carrying out four analyses. To obtain a ten-fold improvement, the formula demands that we carry out a hundred replicates.

The relationship between the standard deviation of the mean and the number of replicates is illustrated in figure 26.

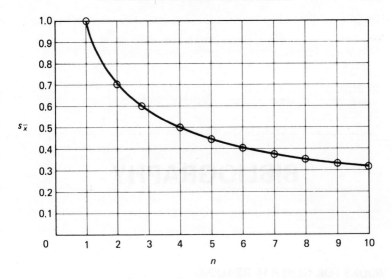

Figure 26   *Relationship between replicate tests n and standard devia-*
tion of the mean (assuming standard deviation of a single
result to be 1.0)

BIBLIOGRAPHY

BOOKS FOR GENERAL READING

Haddock, L. A. (1969). *Analysis in the Chemical Industry*, Pergamon Press, Oxford

Laitinen, H. A. (1960). *Chemical Analysis* (Adv. Chem. Ser.), McGraw-Hill, New York

Stranks, D. R. *et al*. (1966). *Chemistry – A Structural View*, Melbourne University Press

Szabadvary, F. (1966). *History of Analytical Chemistry* (Monographs in Analytical Chemistry), Pergamon Press, Oxford

Wilson, H. N. (1966). *An Approach to Chemical Analysis*, Pergamon Press, Oxford

BOOKS WHICH INCLUDE CHAPTERS ON GENERAL PRINCIPLES

Beaven, G. H. *et al*. (1961). *Molecular Spectroscopy; Methods and Applications in Chemistry*, Heywood, London

Charlot, G. (1964). *Colorimetric Determination of Elements*, Elsevier, Amsterdam

Dean, J. A. (1960). *Flame Photometry* (Adv. Chem. Ser.), McGraw-Hill, New York

Ebdon, L. (1982). *An Introduction to Atomic Absorption Spectroscopy*, Heyden, London

Eckschlager, K. (1969). *Errors, Methods and Results in Chemical Analysis*, Van Nostrand Rheinhold, London

Edisbury, J. R. (1966). *Practical Hints on Absorption Spectrometry*, Hilger, London

Huber, W. (1967). *Titrations in Non-Aqueous Solvents*, Van Nostrand, New York

Jenks, P. and Wall, P. (1980). *Thin Layer Chromatography a Laboratory Introduction*, BDH, Poole

Matlock, G. and Ross Taylor, G. (1961). *pH Measurement and Titration*, Heywood, London

Pattison, J. B. (1969). *A Programmed Introduction to Gas Liquid Chromatography*, Heydon & Son (Gt. Britain)

Sargent, J. R. (1975). *Methods in Zone Electrophoresis*, 3rd edn, BDH, Poole

Skoog, D. A. and West, D. M. (1971). *Principles of Instrumental Analysis*, Holt-Saunders, Philadelphia

SOME USEFUL BOOKS ON GENERAL ANALYTICAL METHODS

'AnalaR' Standards for Laboratory Chemicals, 8th edn, BDH, Poole, 1984

Colorimetric Chemical Analytical Methods, 9th edn, Tintometer, Salisbury, 1980

Sandell, E. B. and Hiroshi Onishi (1978). *Colorimetric Analysis of Traces of Metals*, 4th edn, (Chem. Anal. Ser.), Interscience, New York

Vogel's Textbook of Quantitative Inorganic Analysis, 4th edn, Longman, London and New York, 1978

INDEX